Afroz Alam

Moss Flora of India. An Updated Summary of Taxa

GRIN Publishing

Bibliographic information published by the German National Library:

The German National Library lists this publication in the National Bibliography; detailed bibliographic data are available on the Internet at http://dnb.dnb.de .

Imprint:

Print and binding: Books on Demand GmbH, Norderstedt Germany
ISBN: 978-3-656-92859-1

This book at GRIN:

http://www.grin.com/en/e-book/294363/moss-flora-of-india-an-updated-summary-of-taxa

MOSS FLORA OF INDIA

AFROZ ALAM (Ph.D.)
Associate Professor
Department of Bioscience and Biotechnology
Banasthali University, India

ABOUT THE AUTHOR

DR. AFROZ ALAM is currently working as an Associate Professor in the Department of Bioscience and Biotechnology, Banasthali University, Rajasthan. He obtained his M.Sc. and Ph.D. from University of Lucknow, (U.P.). He also qualified ASRB-NET (ICAR) examination. His areas of specialization are Bryophytes' systematics and bryodiversity. He was awarded J.R.F. and S.R.F. in "All India coordinated Project on Taxonomy Capacity Building for Bryophytes" (AICOPTAX) sponsored by Ministry of Environment & Forests, New Delhi at Bryology Unit, Department of Botany, University of Lucknow during his research. Dr. Alam has collected over 3000 accessions of bryophytes in several plant collection trips to different localities of the country, including some arduous and treacherous terrains of Nilgiri and Palni Hills of Southern Peninsula, Eastern and Western Himalayas and Rajasthan.

Dr. Alam has vast experience of teaching and research in bryology. He has over 50 research publications in prestigious International and National Journals and 4 text books. Dr Alam has also attended and presented papers in a number of seminars. He is a Life member of Association of Plant Taxonomy (APT), India and Indian Bryological Society (IBS). Presently, he is working as an Associate Professor in the Department of Bioscience and Biotechnology, Banasthali University, Rajasthan. He is one of the curators of BURI herbarium. Botanical Survey of India (ENVIS-BSI) approved him as one of the experts of bryology in India.

Dedicated

To

Late Prof. Jan-Peter Frahm

'The bryologist beyond boundaries'

PREFACE

Mosses comprise an imperative component of the mountain forest ecosystem. Mixed climatic conditions have granted India with prosperous moss diversity, yet there is no newer update regarding moss flora for India. The most recent checklist was published in the year 2005 which incorporated 1623 taxa of mosses distributed under 342 genera and 57 families. Ever since, not a single attempt was made for updating Indian moss flora. To fill this lacuna, an effort has been made to accumulate the latest status of Indian mosses in the form of present title ***"Moss Flora of India"***, based on the all accessible and existing reports on mosses so far from the India. It provides the most up-to-date account of floristic diversity in the Indian mosses. This comprehensive listing is supplemented with present status (valid/synonym/doubtful) of each taxon based on *www.theplantlist.org*. Each species have been dealt with as far as achievable complete distributional details inside India. It also incorporates information about their endemism status. A modified classification scheme of Buck and Goffinet (2000) has been used for preparation of this updated checklist. This work would be useful for future studies related to Indian mosses.

The physical map has been adopted from the website of the Department of Tourism, India, while the map of India has been adopted from School Atlas (based on survey of India, 1989). Any deviation from the original map, however, has no significance, political or otherwise.

ACKNOWLEDGEMENTS

The author acknowledges the help rendered by the bryologists from various parts of India for providing research articles related to Indian bryophytes.

Thanks are due to the Prof. Aditya Shastri, Vice Chancellor, Banasthali University, Rajasthan and Prof. Vinay Sharma, Dean, Faculty of Science and Technology, Banasthali University, Rajasthan for providing basic facilities.

Prof. S. C. Srivastava, ex- Head, Department of Botany, University of Lucknow deserves special thanks for their support and encouragement during the course of present study.

Last but not least thanks are due to my family and friends for their continuous support.

I offer my genuine apologies for any inaccuracy and request suggestions from all the readers of this book to improve the publication in succeeding editions.

Afroz Alam

ABOUT THE BOOK

The immense forest cover, sufficient precipitation and vital relative humidity in India provide highly encouraging environment for the lavish prevalence of both terricolous and corticolous mosses. The mosses which occur in specialized habitats like soil surface, soil covered rocks (both moist as well as dry), shaded or exposed rocks, extremely wet rocks are considered as terrestrial while those on other larger trees as corticolous. Majority of substrate receives sufficient precipitation and humidity which provide eco-friendly conditions for the growth and development of these remarkable plants.

The present compilation of the moss flora of India revealed the occurrence of total 1578 species of mosses which belong to 21 orders, under 66 families and 328 genera. Out of these 897 retained their valid status, while 437 species are now considered as a synonym and status of 244 species is still unresolved i.e. doubtful name. 130 taxa have been reported as endemic to India. The updated checklist of mosses of India reveals that the most diversified order is Hypnales with 28 families, followed by Dicranales (6 families); Pottiales and Bryales (4 families each). Order Timmiales, Splachnales and Rhizogoniales have 2 families each and rests of the orders are represented by single families. Whereas, in terms of family, family Pottiaceae is the most prominent having 38 genera followed by Hypnaceae with 20 genera. Genera like *Fissidens* (72 spp.), *Brachythecium* (38 spp.), *Pogonatum* (33 spp.), *Mnium* (25 spp.), *Calymperes* (23 spp.), *Brachymnium* (22 spp.), *Sphagnum* (21 spp.), *Barbula* (21 spp.), *Entodon* (20 spp.) and *Tortula* (15 spp.) are most diversified in India. This great diversity of mosses revealed the potential of India in terms of bryodiversity particularly, the mosses.

***( ) shows number of species**

INTRODUCTION

GEOLOGY, CLIMATE AND VEGETATION OF INDIA

India engages only 2.50% of the overall land of the planet. With a physical area of 329 million ha, it is world's seventh leading country after Russia, Canada, USA, Brazil, Australia and China. India is one of the signatory members of Convention on Biological Diversity (CBD), and also one of the 17 megadiversity countries globally recognized with their geographic extent both on land as well as sea (www.wikipedia.org). It has diverse edaphoclimatic circumstances ranging from the icy mountain climate in the north to a humid tropical one in the south to dry hot desert in the North West (Fig. 1). This mixture of climatic conditions is also reflected in the forest and the biodiversity wealth of the country (Champion and Seth, 1968). Historically, India had over 65% of its total land under forests as early as in 1925 (at present only 20.55%; Ravindranath et al, 2014). The resultant varied climate supplemented with topography created 10 biogeographic zones (Lakshminarayana et al., 2001) namely the Trans-Himalayan, Himalayan, Indian desert, Semi arid, Western Ghats, Deccan peninsula (including the eastern Ghats), Gangetic plains, North-east India, Coasts and Islands comprising their own fragile and unique ecosystem and well defined flora ranking third in Asia and eleventh in the world harboring over 45,000 phytobinomials including lower non-vascular cryptogams, the 'Bryophytes'. India is discernible by diversity of physical features such as peaks, uplands, plains, coastline and isles (Figs. 1 & 2). The brief descriptions of these are given below:

The Himalayas

The Himalayas considered the youngest series of folded mountains on the planet. The Himalayan Region of India is a range that extends ten states of India including Jammu & Kashmir, Himachal Pradesh, Uttarakhand, Arunachal Pradesh and Sikkim, Assam and West Bengal. It forms a curve which is about 2,400 km long. There are three similar ranges in its longitudinal extent (Ahmad, 2011). The region is liable for providing water to a hefty part of the Indian subcontinent. The region physiographically, starting from the foothills of south (Siwaliks) and extends up to Tibetan highland to the north (Trans-Himalaya). The width differs from 150 km in Arunachal Pradesh (eastern) to 400 km in Kashmir (western). The altitudinal deviations are bigger in the eastern part than in the western division. Three major geographical entities, the Himadri (greater Himalaya), Himanchal (lesser Himalaya) and the Siwaliks (outer Himalaya) extending almost continuous throughout its length, are separated by major geological fault lines. Grand rivers like the Indus, Sutlej, Kali, Kosi and Brahmaputra have cut through sharp ravines to run away into the Great Plains. Some of the utmost mountains on this planet are found in this region.

***Trans-Himalayas*:**

This region is the Northern most area in the India extends in the states of Jammu and Kashmir and Himachal Pradesh.

The Himadri:

This is the most constant range of Himalayas (Karakoram Mountains). It holds the highest peaks of the Himalayas. The middling height of the peaks in this range is ca. 6,000 m. The folds of the Great Himalayas are irregular in nature and the center of this part is composed of granite. Because of the supercilious heights, the peaks of this range are snow-bound all through the year.

The Himanchal:

This is the southern part of the Himalayas. The elevation of peaks in this range differs from 3,700 m to 4,500 m. The Middling width of this range is ca. 50 km. This range is mainly composed of extremely compacted and distorted rocks.

The Shiwaliks Himalaya:

This is the farthest range of the southern Himalayas. It lies to the south of the Dhaula Dhar, the average height of this range is 1,500 to 2,000m and the width ranges 10 to 50 km. It includes the Jammu hills and extends to Kangra. On Uttarakhand side, it stretches from Dehra Dun to Almora and finally merge with the southern borders of Nepal. This series is made of uncombined sediments. This can be divided into four sub regions, viz. Kumaon Himalayas, Assam Himalayas, Punjab Himalayas and Nepal Himalayas.

The Indian Desert:

The western part of India has the Great Indian Desert. This is also known as the Thar Desert. It lies towards the western boundaries of the Aravali hills. It is a large (17th largest subtropical desert of the world), arid region in the northwestern part of the Indian subcontinent. In India, it covers about 320,000, of which 60% is in Rajasthan and extends into Gujarat, Punjab, and Haryana. The rainfall is very scanty (<150 mm/year), hence the vegetation of this region is meager.

The Semi Arid Region

The semi arid regions in India encompass largely of the Rann of Kutch and the semi -arid regions of Punjab and Gujarat. The Southern arid regions are in the rain shadow of the Western Ghats covering states of Maharashtra, Karnataka and Tamil Nadu.

The Peninsular India

To the south of the northern plains lies the peninsular plateau. It is triangular in outline. The relief is extremely bumpy. This is the expanse with numerous hill ranges and valleys. Aravali hills, one of the oldest ranges of the world, border it on the north-west side. It is made of the ancient rocks since it was formed from the drifted part of the Gondwana. Expansive and superficial valleys and smoothed hills are the attributes of this plateau. The Vindhyas and the Satpuras are the important ranges. The river Narmada and Tapi flow through these ranges. The upland can be largely alienated into two expanses- the Middle Uplands and the Deccan Plateau.

The Middle Uplands: The Middle Uplands lies to the north of the Narmada waterway. It covers the most important portion of the Malwa highland. The rivers in this region run from southwest to northeast, which points out the slant of this area. The upland also extends eastwards into the Chhotanagpur upland.

The Deccan Plateau: The Deccan plateau of Gondwanaland comprises of one of the most fragile ecosystem "the Western Ghats", one of the 'hot spots' of India, hosting 30% endemic flora and fauna. The Sahyadris or Western Ghats border the plateau in the west and the Eastern Ghats offer the eastern frontier.

The Western and the Eastern Ghats:

The Western Ghats are the chief tropical evergreen region of India, spread over cardinal 220 N to 80 latitude, covering the length of 1400 km. The Western Ghats, known locally as the Sahyadri Hills, are formed by the Malabar Plains and the chain of mountains running parallel to India's western coast, about 30 to 50 km inland. The Western Ghats instigate oceanographic rains as they face the rainy winds from the west. They cover an area of about 160,000 km^2 and stretch for 1,600 kilometers from the country's southern tip to Gujarat in the north, broken up only by the 30 kilometers Palghat Gap.

Unlike, Western Ghats, the Eastern Ghats are broken and jagged. The plateau is loaded with various minerals. The average elevation of Eastern Ghats is about 650m (Alam et al., 2007). The Eastern Ghats, also known as Mahendra Pravata are broken and jagged range of mountains along India's eastern shore. The Eastern Ghats run from West Bengal state in the north, through Odisha and Andhra Pradesh to Tamil Nadu in the south passing some parts of Karnataka. They are wrinkled and cut through by the four major rivers of peninsular India, known as the Mahanadi, Krishna, Godavari and Kaveri.

North East India: The Brahmaputra River indicates the eastern boundary of the Himalayas. Ahead of the Dihang gorge, the Himalayas curve stridently towards south and form the Eastern hills. These hills run all the way through from the north eastern states of India. They are typically composed of sandstones. These hills are identified as Naga Hills, Manipuri Hills, Patkai Hills, and Mizo Hills.

The Gangetic Plains:

The northern Indian plains of India lie to the south of the Himalayas. They are usually flat and level. These are formed by the alluvial deposits laid down by the three rivers- the Indus, the Ganga, the Brahmaputra and their tributaries. These river plains provide fertile land for cultivation. The entire area of the northern plain is about 7 lakh sq. km. It is about 240 to 320 km broad and 2400 km long. The northern plain is divided into three sections, viz. the Ganga Plain, the Brahmaputra Plain and the Punjab Plain.

The Coastal Plains:

The Western Coastal Plains consist of a thin strip of coastal plain ca. 50 km in girth between the west coast of India and the Western Ghats hills, which starts close to the south of Tapi. They are squeezed in between the Western Ghats and the Arabian Sea. The plains initiate at Gujarat in the north and finish at Kerala in the south. It also embraces the states of Maharashtra, Goa and Karnataka. The western coastal plain has two parts. The northern part is known as Northern Circar, while, the southern part is called the Coromandel Coast.

The Eastern Coastal Plains refer to an extensive expand of Indian landmass, lying between the Eastern Ghats and the Bay of Bengal. These plains are wider and level as compared to the western coastal plains. It widens from Tamil Nadu (south) to West Bengal (north). The western coastal plains are very narrow. The eastern coastal plains are much broader. They run along the Arabian Sea on the west and along the Bay of Bengal on the east. There are number of east flowing rivers. The rivers Mahanadi, Godavari, Krishna and Kaveri drain into the Bay of Bengal.

The Islands

Two groups of Islands also form part of India. The Lakshadweep Islands are located in the Arabian Sea. Its area is 32 sq km. This cluster of islands is prosperous in terms of biodiversity. The Andaman and the Nicobar Islands lie to the southeast of the Indian mainland in the Bay of Bengal. They are larger in size and has added a number of islands. These islands also have rich biodiversity.

Prevailing factors of vegetation and flora

Land is one of the most important factors which directly and indirectly governs the vegetation and flora in nature. Varied soil types are vital for the establishment of different types of vegetation and flora.

Climatic conditions like temperature and humidity are the chief factors which govern the nature and level of flora and vegetation of an area. In general, an area with elevated temperature and elevated humidity supports evergreen forest, while an area with elevated temperature and low humidity supports xerophytic vegetation. Likewise, photoperiod (duration of light) is another factor which depends on latitude, altitude, season and duration of the day length. Precipitation also controls the nature and type of vegetation. If an area gets heavy rainfall, it is suitable for the growth of dense vegetation. Conversely, an area with insufficient rainfall is suitable for xerophytes. Ecosystem also has detrimental effects on vegetation. All the biological organisms in an area are mutually dependent on each other. These, along with their physical environment make the ecosystem. A very large ecosystem is called a biome which is identified on the basis of plant communities (www.envindia.com).

NATURAL VEGETATION

Natural vegetations are offerings of nature. They grow naturally. They follow the climatic variables. Because of a mixture of climates, a wide range of natural vegetation grows in India. Types of natural vegetation vary according to climate, soil and altitude. The naturally growing plants cover without human any interference or help is called natural vegetation. They have their characteristic flora (www.ncert.nic.in).

Types of Natural Vegetation

The following are the principal types of natural vegetation in India (Kishwan et al.,2009):

(1) Tropical Rain Forests

These forests grow in areas where precipitation is > 200 cm/year. They are mainly found on the slants of the Western Ghats and the north-eastern regions of Arunachal Pradesh, Meghalaya, Assam, Nagaland, the foothill areas of the Himalayas and the Andaman groups of Islands. Vegetation in such a forest has a multilayered structure. Sisthu, chaplash, bamboos, garjan, ebony, mahogany, rosewood, rubber, cinchona and sandalwood are some of the commercially important trees of these rainforests.

(2) Tropical Deciduous or Monsoon Forests

In India these forests are the most prevalent also known as Monsoon forests. They are common in those regions which acquire annual precipitation between 70 cm and 200 cm/year. The trees are deciduous which shed their leaves during summer.

On the basis of availability of water these can be divided into following two types.

(a) Moist Deciduous Forest: The moist deciduous forests need annual rainfall between 100 cm and 200 cm. Northeastern states hold these forest. However, they are also found on the eastern slopes of the Western Ghats, the foothills of the Himalayas, Jharkhand, West Orissa and Chhattisgarh.

(b) Dry Deciduous Forest: The dry deciduous forests grow in areas where the annual rainfall is between 70 cm and 100 cm. These are found in areas of the central Deccan plateau, southeast of Rajasthan, Punjab, Haryana and parts of Uttar Pradesh and Madhya Pradesh. Most of these areas are used for agriculture.

(c) Semi-deserts and Desert vegetation

These forests exist in the regions that receive less than 70 cm of annual rainfall. North-western parts of India (Rajasthan and parts of Gujarat, Punjab aand Karnataka) hold this type of vegetation. The main plant species in such a forest are xerophytes.

(d) Montane Forests

In India, Montane forests are classified into three categories; the montane wet temperate forests, Himalayan moist temperate forests and Himalayan dry temperate forests. These forests differ significantly according to altitude with varying rainfall and temperature along the slopes of the mountain.

(i) The montane wet temperate forests include shola forests. These are found between a height of 1000 and 2000 m. Evergreen broad-leaf conifers are abounding in these forests.

(ii) The Eastern Himalayan forest is an eco-region of temperate broadleaf forest. These are found in the middle elevations (1500 -3000m) of the eastern Himalayas. Gymnospermous trees of family Coniferaceae flourish in such forests.

(iii) Alpine plants grow in the alpine climate, which occurs at high altitude (>3600m) and above the tree line. Alpine plants grow mutually as a plant community. Alpine plants are limited to a single taxon. Rather, many diverse plant species live in this environment. These plants must adapt to the harsh conditions of the alpine environment, which include low temperatures, dryness, ultraviolet radiation, and a short growing season. These forests are mainly found along the southern slopes of the Himalayas and at high altitudes in southern and north-eastern India.

(e) *Mangrove Forests*

Mangrove forests are found in the deltas of the Ganga, the Mahanadi, the Krishna, the Godavarai and the Kaveri. They are also known as 'Tidal Forests'; because their dense growth depends upon tidal water, which submerges the deltaic lands during high tides (www.envindia.com; www.importantindia.com).

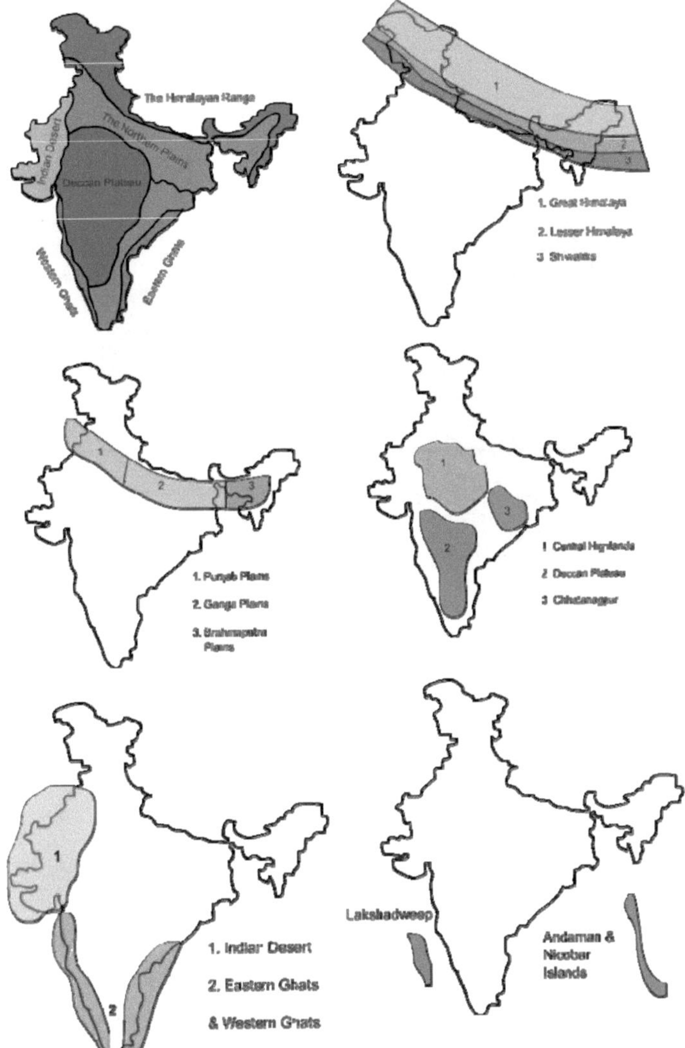

Figure 1: Major Physiographic Divisions of India (Oxford, 2014)

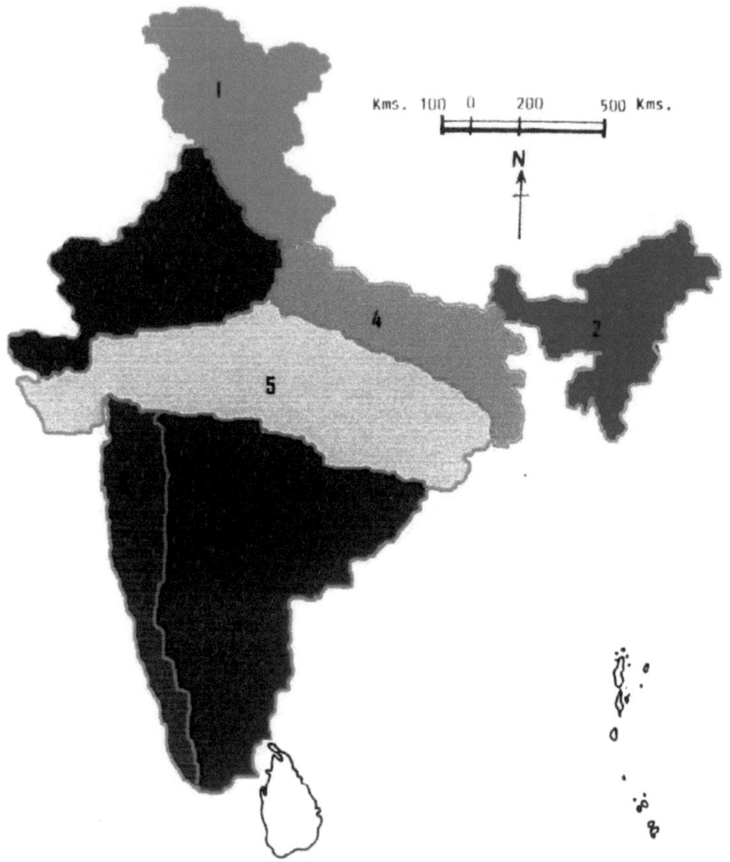

Figure 2. Bryogeographical regions of India: (1) West Himalayan region, (2) East Himalayan region, (3) Punjab and west Rajasthan, (4) Gangetic plains, (5) Central India, (6) west Coast, (7) East coast and Deccan plateau (Lal, 2005)

BRYOFLORISTIC WEALTH

After angiosperms, bryophytes constitute the second most varied land plant group. There are three major bryophyte phyla: Hepatophyta the "liverworts" approx. 6000-8000 species, Anthocerophyta the "hornworts" approx. 150 species, Bryophyta the "mosses" includes about 680 genera and approx. 10000-12000 species. Bryophytes are characterized by an alternation of generations (heteromorphic) with the haploid generation (gametophyte) being the dominant phase of the life cycle. They are also referred as "Lilliputians of Plant Kingdom" due to their minuscule size. They have a dichotomous branching pattern, a primitive type a branching. They do not have any vascular supply in the midrib. They are habitually known as 'Amphibians of the plant kingdom' as they need water to complete the process of fertilization (Glime, 2007).

Mosses

As mentioned earlier, mosses have utmost species diversification. They habitually grow in moist and shady places as crowded greenish cluster or carpet. The plants are typically composed of simple, unistartose leaves, covering a thin axis that supports them. Plants are atracheate, as they do not have vascular tissue. At maturity, they produce sporophytic structures with massive capsules containing spores. They are usually 1–12 cm tall, however, some species like, *Dawsonia* are much larger, which can grow to 50 cm in length and considered as the tallest moss in the world (Schofield, 1985).

Mosses belong to phylum (division) Bryophyta, which previously also included liverworts and hornworts. Now liverworts and hornworts have separate divisions. There are roughly 10,000-12,000 taxa of moss exist under the Bryophyta (Vitt, 1984). The foremost viable use of mosses is for ornamental purposes, such as in gardens and in the florist trade. Conventional uses of mosses included as filling and as they have great capability to soak up liquids up to 20 times their weight (*Sphagnum* can absorbs 200 times) they are extensively used in nursery practices (Glime, 2007).

Life Cycle

The life-cycle of moss exhibits a heteromorphic alternation of generations. It has two very distinct phases, the haploid or gametophytic and the diploid or sporophytic phases. The gametophytic phase is the dominant phase that commences with germination of haploid spore which considered as the first cell of gametophyte. The gametophyte has two growth stages: (i) Protonema stage, which is the juvenile stage represented by prostrate, green and branched thread-like structure. This is an ephemeral stage in the life cycle of a moss; and (ii) leafy stage or gametophores, which is an erect cylindrical shoot with persistent leaves and sex organs. Usually, a solitary carpet of protonema may develop several shoots, ensuing in an aggregation of moss. The main axis can be branched or unbranched. The branches always arise below the leaves. The leaves are simple, minute sessile and usually are one cell in thickness (Shaw and Goffinet, 2000). There are two mode of reproduction:

(i) Vegetative reproduction: It takes place by fragmentation, stolons, branching of protonema, special leafy shoots, gemmae and persistent apices.

(ii) Sexual reproduction: From the apices of main stem and branches of the gametophores development of sex organs in groups takes place. The plants are dioicous or monoicous. In dioicous species, male and female sex organs are borne on different plants (Glime and Knoop, 1986). In monoicous (also known as autoicous) species, both the sex organs are borne on the same plant. The female sex organs are called archegonia, which are protected by a group of modified protective leaves known as the perichaetum. The archegonia are stalked with a much elongated neck and massive venter. These are flask-shaped structure with an open neck through which the male sperm swim towards the venter. The male sex organs are termed as antheridia, these are club-shaped, narrow and elongated. The antheridial jacket is single layered. These are enclosed by modified protective leaves called the perigonium. The surrounding leaves of atheridium form a cup like cavity, allowing the sperm contained in the cavity to be splashed to adjacent stalks (female branches) by falling rain drops, for this reason this is called 'Splash cup mechanism' (Glime, 2007).

Fertilization takes place with the aid of water. Water confers the swimming movement of male gametes from the antheridium towards the archegonium to complete the process of fertilization. The male gametes of mosses are biflagellate structures and exhibits chemotactic movement. After fertilization, the juvenile sporophyte pushes its way out of the archegonial venter. After six months for the sporophyte becomes mature. The sporophyte is usually differentiated into foot, seta and capsule. The elongated seat raises the capsule much above the gametophyte to facilitate the dispersal off spores. The capsule is covered by a cap like structure called the operculum. The capsule and operculum both are enclosed by an additional structure called calyptra which is the remnants of the archegonial venter hence haploid in nature. It normally falls off from full-grown capsule. The capsule usually has a peristome which helps in the dispersal of spores. Within the capsule, sporogenous mass undergo reduction division and form haploid spores. Most mosses depend on the wind for their dispersal.

It has recently been found that few insects can affect fertilization process of moss because they attract towards the specific moss emitted odour. For example, fire moss, releases unusual and composite volatile organic odour. Female plants give off added compounds than male plants. Springtails prefer usually the female plants of this taxa, and increase moss fertilization, suggesting a odour-mediated relationship equivalent to the plant-pollinator relationship found in many seed plants. Similarly, *Splachnum sphaericum* increases insect pollination at higher rate by magnetizing flies to its sporangia with a intense odour of carrion. Bright red coloured inflated collars below each spore filled capsule serve as a strong visual hint for pollinators. Flies fascinated to the moss transmit its spores to fresh droppings of cattle, which is the conducive habitat of this taxa (Glime, 2007).

Habitat

They are better adapted to terrestrial habitats than other groups of bryophytes. Since moss gametophytes have no water conducting tissue through the plant or water holding systems to avoid tissue water from evaporation, mosses need a moist environment in which they grow, and availability of fluid water for reproduction. Since mosses are autotrophic (except saprophytic genus *Buxbaumia*) they need sufficient sunlight to conduct photosynthesis. Selection of substrate differs according to the specis. They grow on moist soil, rocks, damp walls, old buildings and on the stem of trees in tropical forest. Moss species growing on or in the shade big trees are often species specific such as preferring reduced leaf tree like pinus to broadleaf trees like eucalyptus. The remarkable point about the mosses is their mode of nutrition, they always grow on trees as epiphytes, they never adopt the parasitic mode on the phorophyte.

In general, mosses can grow in a variety of habitats. For example, *Barbula comosa, Weisia endultula, Brachymenium walkeri* grow on dry faces of cliffs of gneiss and granites; *Bryum giganteum*, *Hypophila comosa* and *Barbula indica* on pegmatites, lime and black loam; *Fissidens lutescens, F. walkeri, Trematodon ceylonesis* and *Bryum wiohlii* on banks of stream in shady places; *Bryum ramosum* and *B. doliolum* on dry banks; *Leucoloma walkeri, Fissidens aomalous* and *Bryum apalodictyoides* on dead wood and decaying tree trunks: *Leucoloma renauldii, Tortella hyalinoblasta* and *Macromitrium sulcatum* on large trees forming felts (in dense jungles) etc. are found growing in various parts of the country. Several mosses such as *Polytrichum* are reported to grow on rocks in extremely xerophytic conditions, sometimes in association with lichens. These help in succession of the vegetation (xerosere). Likewise, *Porella platyphylloidea* growing on rocks can withstand without water for several months. It shows extreme drought resistance.

One of the ecological significance of bryophytes is to avoid erosion by holding the soil particles trough carpet form of growth and retaining the greater amount of water. They also slow down the rapid runoff water and melted snow. These also provide a soft bed to the seeds and moist conditions favourable to their growth. Thus, yet insignificant, bryophytes play a quiet but fairly vital role in the nature (Proctor, 1984).

DISTRIBUTION OF MOSSES IN INDIA

India is one of the 17 mega biodiversity countries in the world and one of the hot spots of biodiversity.. The large area and a variety of phytoclimatic conditions contribute to the great diversity of the Indian flora (Singh, 1997; 2001; Alam et al., 2007). In case of bryophytes and especially mosses, rich diversity is found in different regions of India. According to a rough estimate about 2000 species of mosses, 816 species of liverworts and 34 species of hornworts are occurring in India. The plants are distributed in Eastern and Western Himalayas, South India, Rajasthan, Gujarat, Punjab, Central India, Andaman and Nicobar Islands (Fig. 2). According to Lal (2005) about 2480 taxa of bryophytes (including intraspecific taxa) are reported from India (including island groups, and Sikkim), comprising about 722 taxa of liverworts in 128 genera and 52 families, 36 taxa in 6 genera and 2 families of hornworts and about 1623 taxa in 342 genera and 57 families of mosses.

Several bryologists of India have assessed the moss flora of different bryological regions from time to time, such as Gangulee (1969-80) published “Mosses of Eastern India and Adjacent regions” in eight fascicles which included 990 species. Chopra (1975) dealt with nearly 2,000 species belonging to 329 genera under 56 families. He listed many species as "likely to occur" and also *nomen nuda*. Lal (2005) published "A checklist of Indian mosses" and listed 1623 taxa of mosses from India. After that, no valid attempt has been made to provide a complete wealth of mosses in India with their current status. This compilation work is an attempt to provide current status of the moss flora of India which includes mosses of Western Himalayas, Eastern Himalayas, South India, Central India, Gangetic plains, Punjab, Rajasthan, Gujarat Andaman and Nicobar Islands (Figure 1 & 2).

The assets of this work include almost all previous work and reports from these regions of India, like, Dixon, 1909; Dabhade, 1969; Chopra, 1975; Vohra, 1970; Gangulee, 1969-1980; Chopra and Kumar, 1981; Tewari and Pant, 1994; Deora and Chaudhary, 1996; Dabhade, 1998; Madhusoodanan and Nair, 2004; Lal, 2005; Saxena and Gangwar, 2005; Nair and Madhusoodanan, 2006; Saxena et al., 2006; Madhusoodanan et al., 2007; Manju et al., 2007; Saxena et al, 2007; Nath et al, 2007; Daniels and Daniel., 2007; Kuamar and Krishnamurthy, 2007; Nath et al, 2008; Aziz and Vohra, 2008; Saxena and Arfeen, 2009; Saxena et al, 2010; Singh et al., 2010; Daniels, 2010; Dandotiya et al., 2011; Manju et al., 2011; Alam et al, 2012; Verma et al., 2011; Alam, 2013a,b; Daniels and Kariyappa, 2013; Asthana and Sahu, 2013, Rajesh et al.; 2013; Schwarz, 2013; Schwarz and Frahm, 2014; Rajesh and Manju, 2014; Alam et al., 2014). This work aims to be all-inclusive for species of mosses. It contains a total 757 moss species along with their upper hierarchy. The list has three categories for the species names viz. Valid name, synonym and doubtful name. These categories are based on available literature from reliable resources like 'The Plant List' (2010) which was prepared in collaboration of the Royal Botanic Gardens, Kew and Missouri Botanical Garden by combining multiple checklist datasets held by these institutions and other collaborators. This present list also follows this and provides the valid name, synonyms and also provides doubtful name for which the contributing data sources did not contain sufficient evidence to decide whether they were accepted or synonyms.

Therefore, this assemblage provides comprehensive, reorganized and an updated account on the moss flora of India.

METHODOLOGY

This study is fundamentally based on all previous and recent reports regarding moss flora of India. The orders are arranged alphabetically for feasibility. All moss species included in the list were checked against the TROPICOS database (at the Missouri Botanical Garden). A modified classification scheme of Buck and Goffinet (2000) has been used for preparation of this updated checklist.

THE CHECKLIST OF MOSSES

A. ORDER: TAKAKIALES

Family: Takakiaceae S. Hatt. & Inoue.

1. *Takakia* S. Hatt. & Inoue.

1. *Takakia ceratophylla* (Mitt.) Grolle; **Present status: Valid name**

Distribution in India: Eastern Himalayas

B. ORDER: ANDREAEALES LIMPR.

Family: Andreaeaceae Dum.

2. *Andreaea* Hedw.

2.*Andreaea commutata* Müll. Hal.; Present status: **Synonym of** *Andreaea rupestris* var. *fauriei* (Besch.) Takaki

Distribution in India: Eastern Himalayas (Endemic to India)

3. *Andreaea densifolia* Mitt.; **Present status: Valid name**

Distribution in India: Eastern Himalayas (Endemic to India)

4. *Andreaea indica* Mitt.; Present status: **Doubtful**

Distribution in India: Eastern Himalayas (Endemic to India)

5. *Andreaea rigida* Wilson in Mitten; **Present status: Valid name**

Distribution in India: Eastern Himalayas (Endemic to India)

6. *Andreaea kashyapii* Dixon & Vohra & Wadhwa Endemic to India; Present status: **Synonym of** *Didymodon subandreaeoides* (Kindb.) R.H. Zander

Distribution in India: Endemic to India to Western Himalyas.

7. *Andreaea rupestris* Hedw.; **Present status: Valid name**

Distribution in India: Western Himalayas and Eastern Himalayas

C. ORDER:ARCHIDIALES LIMPR.

Family: Archidiaceae Schimp.

3. ***Archidium*** Brid.

8. *Archidium birmannicum* Mitt. ex Dixon; **Present status: Valid name**

Distribution in India: Gangetic plains and South India

9. *Archidium indicum* Müll. Hal.; **Present status:** Doubtful name

Distribution in India: Western Himalayas and Central India.

10. *Archidium microthecium* Dixon & P. de la Varde; **Present status:** Doubtful name

Distribution in India: South India.

11. *Archidium octosporum* Dixon & P. de la Varde; **Present status: Synonym of** *Archidium ohioense* Schimp. ex Müll. Hal.

Distribution in India: South India

D. ORDER:BRYALES LIMPR.

Family: Bryaceae Schwägr.

4. ***Anomobryum*** Schimp.

12. *Anomobryum astorense* (Broth.) Broth. ; **Present status: Valid name**

Distribution in India: Western Himalayas

13. *Anomobryum auratum* (Mitt.) A. Jaeger.; **Present status: Valid name**

Distribution in India: Western Himalayas, Eastern Himalayas, South India

14. *Anomobryum brachymenioides* Dixon & P. de la; **Present status: Synonym of** *Bryum brachymenioides* (Dixon & P. de la Varde) Ochi

Distribution in India: South India

15. *Anomobryum cymbifolium* (Lindb.) Broth.; **Present status: Valid name**

Distribution in India: Western Himalayas, Eastern Himalayas, South India

16. *Anomobryum filiforme* (Griff.) A. Jaeger ; **Present status: Valid name**

Distribution in India: Western Himalayas, Eastern Himalayas, South India

17. *Anomobryum filiforme* var. *concinnatum* (Spruce) Loesk. (Asthana & Sahu, 2013); **Present status: Valid name**

Distribution in India: Western Himalayas, Eastern Himalayas, South India

18. *A. kashmirense* (Broth.) Broth.; **Present status: Valid name**

Distribution in India: Western Himalayas

19. *Anomobryum latifolium* Cardot & P. de la Varde; **Present status:** Synonym of *Bryum auratum* Mitt.

Distribution in India: South India

20.*Anomobryum marginatum* Dixon & Badhw.; Present status: **Synonym of** *Bryum blandum* subsp. *handelii* (Broth.) Ochi

Distribution in India: Western Himalayas

21. *Anomobryum nitidum* (Mitt.) A. Jaeger.; **Present status: Valid name**

Distribution in India: Western Himalayas and Eastern Himalayas

22. *Anomobryum pellucidum* Dixon & Badhw.; Present status: **Synonym of** *Bryum himalayanopellucidum* Ochi

Distribution in India: Western Himalayas

23. *Anomobryum schmidii* (Müll. Hal.) A. Jaeger; **Present status: Valid name**

Distribution in India: South India

24. *Anomobryum subnitidum* Cardot & P. de la Varde; **Present status: Doubtful**

Distribution in India: South India

5. *Brachymenium* Schwägr.

25. *Brachymenium acuminatum* Harv.; **Present status: Valid name**

Distribution in India: Eastern Himalayas and South India

26. *Brachymenium alpinum* Ochi; **Present status: Valid name**

Distribution in India: Eastern Himalayas

27. *Brachymenium bryoides* Hook. & Schwägr.; **Present status: Valid name**

Distribution in India: Western Himalayas, Eastern Himalayas, South India

28. *Brachymenium clavariaeforme* (Müll. Hal.) A. Jaeger; **Present status: Valid name**

Distribution in India: South India

29. *Brachymenium cristatum* (Müll. Hal.) A. Jaeger; **Present status: Valid name**

Distribution in India: South India

30. *Brachymenium extenuatum* (Mitt.) A. Jaeger; **Present status: Valid name**

Distribution in India: South India

31. *Brachymenium exile* (Dozy & Molk.) Bosch & Sande Lac.; **Present status: Valid name**

Distribution in India: Western Himalayas, Eastern Himalayas, South India

32. *Brachymenium fischeri* Cardot & Dixon; **Present status: Doubtful**

Distribution in India: South India

33. *Brachymenium flaccidisetum* (Müll. Hal.) A. Jaeger; **Present status: Valid name**

Distribution in India: South India

34. *Brachymenium indicum* (Dozy & Molk.) Bosch & Sande Lac.; **Present status: Valid name**

Distribution in India: Gangetic plains

35. *Brachymenium lanceolatum* Hook. & Wilson; **Present status: Valid name**

Distribution in India: Central India

36. *Brachymenium leptostomoides* (Müll. Hal.) A. Jaeger; **Present status: Doubtful**

Distribution in India: South India

37. *Brachymenium longicolle* Thér.; Present status: **Synonym of** *Brachymenium leptophyllum* (Bruch & Schimp. ex Müll. Hal.) Bruch & Schimp. ex A. Jaeger

Distribution in India: Eastern Himalayas

38. *Brachymenium longidens* Renauld & Cardot; **Presnt status:** Doubtful name

Distribution in India: Eastern Himalayas

39. *Brachymenium microstomum* Harv.; **Present status: Synonym of** *Pseudopohlia bulbifera* R.S. Williams

Distribution in India: Eastern Himalayas and Central India

40. *Brachymenium nepalense* Hook.; **Present status: Valid name**

Distribution in India: Western Himalayas, Eastern Himalayas, South India

41. *Brachymenium ochianum* Gangulee; **Present status: Synonym of** *Brachymenium capitulatum* (Mitt.) Paris

Distribution in India: Eastern Himalayas

42. *Brachymenium ptychothecium* (Besch.) Ochi; **Present status: Valid name**

Distribution in India: Eastern Himalayas

43. *Brachymenium rugosum* (Müll. Hal.) A. Jaeger; **Present status: Valid name**

Distribution in India: South India

44. *Brachymenium sikkimense* Renauld & Cardot; **Present status: Doubtful name**

Distribution in India: Eastern Himalayas

45. *Brachymenium velutinum* (Müll. Hal.) A. Jaeger; **Present status: Valid name**

Distribution in India: South India

46. *Brachymenium walkeri* Broth.; **Present status: Doubtful name**

Distribution in India: Eastern Himalayas and south India

6. *Bryum* Hedw.

47. *Bryum allionii* Broth.; Present status: **Synonym of** *Bryum mildeanum* Jur.

Distribution in India: Western Himalayas, Eastern Himalayas, South India

48. *Bryum alpinum* Huds. & With. ; **Present status: Valid name**

Distribution in India: Western Himalayas, Eastern Himalayas, South India

49. *Bryum alpinum* var. *mildeanum* (Jur.) Podp.; Present status: **Synonym of** *Bryum mildeanum* Jur.

Distribution in India: Western Himalayas, Eastern Himalayas, South India

50. *Bryum apiculata* Schwägr.; Present status: **Synonym of** *Bryum mildeanum* Jur.

Distribution in India: Western Himalayas, Eastern Himalayas, South India and Gangatic Plains

51. *Bryum ambiguum* Duby; **Present status: Valid name**

Distribution in India: South India

52. *Bryum andrei Cardot & P. de la Varde*; **Present status:** Synonym of *Bryum paradoxum* Schwägr

Distribution in India: South India

53. *Bryum argenteum* Hedw.; **Present status: Valid name**

Distribution in India: Western Himalayas, Eastern Himalayas, South India, Central India and Rajasthan

54. *Bryum argenteum* var. *lanthanum* (P. Beauv.) Hampe; **Present status: Synonym of** *Bryum argenteum* Hedw.

Distribution in India: Western Himalayas, Eastern Himalayas, South India and Central India

55. *Bryum. atrovirens* Brid.; **Present status: Valid name**

Distribution in India: Western Himalayas and, Eastern Himalayas

56. *Bryum badhwarii* Ochi ; Present status: **Synonym of** *Bryum kashmirense* Broth.

Distribution in India: Western Himalayas and Eastern Himalayas

57. *Bryum bessonii* Renauld & Cardot; **Present status: Valid name**

Distribution in India: South India

58. *Bryum bicolor* Dicks.; Present status: **Synonym of** *Bryum dichotomum* Hedw.

Distribution in India: Western Himalayas, Eastern Himalayas and Rajasthan

59. *Bryum billardieri* Schwägr. ; **Present status: Valid name**

Distribution in India: South India

60. *Bryum bornholmense* Wink. & R. Ruthe; **Present status: Valid name**

Distribution in India: Rajasthan

61. *Bryum bryoides* (R. Br.) Ångström ; Present status: **Synonym of** *Bryum arcticum* (R. Br.) Bruch & Schimp.

Distribution in India: Western Himalayas

62. *Bryum caespiticium* Hedw.; **Present status: Valid name**

Distribution in India: Western Himalayas and Eastern Himalayas

63. *Bryum capillare* Hedw.; Present status: **Synonym of** *Ptychostomum capillare* (Hedw.) D. T. Holyoak & N. Pedersen

Distribution in India: Western Himalayas, Eastern Himalayas and, South India

64. *Bryum cellular* Hook.; **Present status: Valid name**

Distribution in India: Western Himalayas, Eastern Himalayas and Nicobar Island.

65. *Bryum. coronatum* Schwägr.; **Present status: Valid name**

Distribution in India: Western Himalayas, Eastern Himalayas, South India, Central India, Rajasthan, Gangetic plains and Andaman Island.

66. *Bryum doliolum* Duby, **Present status:** Synonym of *Bryum coronatum* var. *doliolum* (Duby) A. Jaeger

Distribution in India: South India

67. *Bryum euryphyllum* Dixon & P. de la Varde; **Present status: Valid name**

Distribution in India: South India

68. *Bryum flaccum* Wilson ex Mitt.; **Present status: Doubtful name**

Distribution in India: Eastern Himalayas

69. *Bryum formosum* **Mitt. ; Present status: Synonym of** *Bryum wightii* Mitt.

Distribution in India: South India

70. *Bryum ghatense* Broth. & Dixon; **Present status:** Synonym of *Bryum mildeanum* Jur.

Distribution in India: South India

71. *Bryum ghatense* var. *satarense* Broth. & Dixon; **Present status:** Synonym of *Bryum mildeanum* Jur.

Distribution in India: South India

72. *Bryum heterophyllum* Warnst.; **Present status: Synonym of** *Bryum dichotomum* Hedw.

Distribution in India: South India

73. *Bryum klinggraeffii* Schimp.; **Present status: Valid name**

Distribution in India: Rajasthan and Gangetic plains

74. *Bryum lamprostegum* Müll. Hal.; ***Present* status: Doubtful**

Distribution in India: South India

75. *Bryum medianum* Mitt. ; **Present status**: **Synonym of** *Bryum neelgheriense* var. *neelghreniense*

Distribution in India: Western Himalayas, Eastern Himalayas, South India

76. *Bryum montagneanum* **Müll. Hal.; Present status**: Synonym of *Brachymenium pendulum* Mont.

Distribution in India: South India

77. *Bryum muehlenbeckii* Bruch & Schimp.; **Present status: Valid name**

Distribution in India: Western Himalayas, Eastern Himalayas, South India (Endemic to India)

78. *Bryum nitens* Harv **Present status**: **Synonym of** of *Gemmabryum apiculatum* (Schwägr.) J.R. Spence & H.P. Ramsay

Distribution in India: South India

79. *Bryum pachycladum* Cardot ex P. de la Varde; **Present status: Doubtful**

Distribution in India: South India

80. *Bryum paradoxum* Schwägr.; **Present status: Valid name**

Distribution in India: Western Himalayas, Eastern Himalayas, South India

81. *Bryum paradoxum* var. *reflexifolium* (Ochi) Ochi; Present status: **Synonym of** *Bryum reflexifolium* (Ochi) Ochi

Distribution in India: Eastern Himalayas

82. *Bryum plumosum* Dozy & Molk.; Present status: **Synonym of** *Gemmabryum apiculatum* (Schwägr.) J.R. Spence & H.P. Ramsay

Distribution in India: Western Himalayas, Eastern Himalayas, South India and Gangetic Plains

83. *Bryum porphyroneuron* subsp. *erythropus* M. Fleisch.; Present status: **Synonym of** *Bryum mildeanum* Jur.

Distribution in India: Western Himalayas, Eastern Himalayas, South India

84. *Bryum pseudotriquetrum* (Hedw.) P. Gaertn., B. Mey. & Scherb.; **Present status: Valid name**

Distribution in India: Western Himalayas, Eastern Himalayas

85. *Bryum ramosum* (Harv.) Mitt.; Present status: **Synonym of** *Bryum billarderi* Schwägr

Distribution in India: South India

86. *Bryum ramosum* subsp. *zollingeri* (Duby) M. Fleisch.; **Present status: Doubtful name**

Distribution in India: South India

87. *Bryum. recurvatum* Broth.; Present status: **Synonym of** *Bryum recurvulum* Mitt.

Distribution in India: South India

88. *Bryum retusifolium* Cardot & P. de la Varde; **Present status: Valid name**

Distribution in India: South India

89. *Bryum rubens* Mitt.; **Present status: Valid name**

Distribution in India: Rajasthan

90. *Bryum salakense* Cardot; **Present status: Valid name**

Distribution in India: India Orientalis

91. *Bryum tuberosum* Mohamed & Damanhuri; **Present Status: Synonym of** *Rosulabryum tuberosum* (Mohamed & Damanhuri) J.R. Spence

Distribution in India: South India

92. *Bryum turbinatum* (Hedw.) Turner ; **Present status: Valid name**

Distribution in India: Western Himalayas

93. *Bryum. uliginosum* (Brid.) Bruch & Schimp.; **Present status: Valid name**

Distribution in India: Western Himalayas, Eastern Himalayas

94. *Bryum vellei* Cardot & P. de la Varde; **Present Status: Synonym of** *Rosulabryum billarderi* (Schwägr.) J.R. Spence

Distribution in India: South India

95. *Bryum vellei var. robustum* Dixon & P. de la Varde; *Present* **Status: Synonym of** *Rosulabryum billarderi* (Schwägr.) J.R. Spence

Distribution in India: South India

96. *Bryum weigelii* Spreng.; **Present status: Valid name**

Distribution in India: Western Himalayas

97. *Bryum wightii* Mitt.; **Present status: Valid name**

Distribution in India: South India

7. *Mielichhobryum* J. P. Srivast.

98. *Mielichhobryum sahayadrense* J.P. Srivast. ; **Present status**: **Doubtful name**

Distribution in India: Western Himalayas and South India

8. *Mniobryum* Limpr.

99. *Mniobryum delicatulum* (Hedw.) Dixon ; Present status: **Synonym of** *Pohlia melanodon* (Brid.) A.J. Shaw

Distribution in India: Western Himalayas

100. *Mniobryum ludwigii* (Spreng. & Schwägr.) Loeske; **Present status: Synonym of** *Pohlia ludwigii*

Distribution in India: Western Himalayas, Eastern Himalayas

101. *Mniobryum wahlenbergii* (F. Weber & D. Mohr) Jenn.; **Present status: Valid name**

Distribution in India: Western Himalayas, Eastern Himalayas

9. *Plagiobryum* Lindb.

102. *Plagiobryum demissum* (Hook.) Lindb.; **Present status: Valid name**

Distribution in India: Eastern Himalayas

103. *Plagiobryum zierii* (Dicks. ex Hedw.) Lindb.; **Present status: Valid name**

Distribution in India: Eastern Himalayas

10. *Rhodobryum* (Schimp.) Hampe

104. *Rhodobryum giganteum* (Schwägr.) Paris ; **Present status: Valid name**

Distribution in India: Western Himalayas, Eastern Himalayas and South India

105. *Rhodobryum laxelimbatum* (Hampe ex Ochi) Z. Iwats. & T.J. Kop.; **Present status: Valid name**

Distribution in India: Eastern Himalayas

106. *Rhodobryum madurense* Dixon & P. de la Varde; **Present status:** Synonym of *Rhodobryum commersonii* (Schwägr.) Paris

Distribution in India: South India

107. *Rhodobryum ontariense* (Kindb.) Paris ; **Present status: Valid name**

Distribution in India: Eastern Himalayas

108. *Rhodobryum roseum* (Hedw.) Limpr.; **Present status: Valid name**

Distribution in India: Western Himalayas, Eastern Himalayas

Family: Mniaceae Schwägr.

9. *Epipterygium* Lindb.

109. *Epipterygium tozeri* (Grev.) Lindb.; **Present status: Valid name**

Distribution in India: Western Himalayas, Eastern Himalayas

12. *Leptobryum* (Bruch & Schimp.) Wilson

110. *Leptobryum pyriforme* (Hedw.) Wilson ; **Present status: Valid name**

Distribution in India: Western Himalayas, Eastern Himalayas

13. *Mielichhoferia* Nees & Hornsch.

111. *Mielichhoferia acutifolia* Broth.; Present status: **Doubtful name**

Distribution in India: Western Himalayas

112. *Mielichhoferia assamica* Dixon; **Present status: Valid name**

Distribution in India: Eastern Himalayas (Endemic to India)

113. *Mielichhoferia badhwarii* Dixon ; Present status: **Doubtful name**

Distribution in India: Western Himalayas

114. *Mielichhoferia himalayana* Mitt.; **Present status: Valid name**

Distribution in India: Western Himalayas

115. *Mielichhoferia lahulensis* R. S. Williams ; Present status: **Doubtful name**

Distribution in India: Western Himalayas

116. *Mielichhoferia pilifera* E. B. Bartram ; Present status: **Doubtful name**

Distribution in India: Western Himalayas

117. *Mielichhoferia schmidii* Müll. Hal.; Present status: **Doubtful name**

Distribution in India: South India

14. *Mnium* Hedw.

118. *Mnium affine* Blandow & Funck ; Present status: **Synonym of** *Plagiomnium affine* (Blandow & Funck) T.J. Kop.

Distribution in India: Western Himalayas

119. *Mnium confertidens* (Lindb. & Arnell) Kindb.; Present status: **Synonym of** *Plagiomnium confertidens* (Lindb. & Arnell) T.J. Kop.

Distribution in India: Eastern Himalayas

120. *Mnium cuspidatum* Neck. & Lindb.; Present status: **Synonym of** *Plagiomnium affine* (Blandow & Funck) T.J. Kop.

Distribution in India: Western Himalayas, Eastern Himalayas

121. *Mnium heterophyllum* (Hook.) Schwägr. ; **Present status: Valid name**

Distribution in India: Western Himalayas, Eastern Himalayas

122. *Mnium integrum* Bosch & Sande Lac.; Present status: **Synonym of** *Plagiomnium integrum* (Bosch & Sande Lac.) T.J. Kop.

Distribution in India: Western Himalayas, Eastern Himalayas

123. *Mnium japonicum* Schimp.; Present status: **Synonym of** *Plagiomnium acutum* (Lindb.) T.J. Kop.

Distribution in India: Western Himalayas, Eastern Himalayas

124. *Mnium kumaonensis* Srivastava ; Present status: **Doubtful name**

Distribution in India: Western Himalayas

125. *Mnium laevinerve* Cardot ; **Present status: Valid name**

Distribution in India: Western Himalayas, Eastern Himalayas

126. *Mnium lycopodioides* Schwägr.; Present status: **Synonym of** *Mnium ambiguum* H. Müll.

Distribution in India: Western Himalayas, Eastern Himalayas

127. *Mnium marginatulum* Müll. Hal.; Present status: **Synonym of** *Pohlia mauiensis* (Broth. & E.B. Bartram) W. Schultze-Motel

Distribution in India: Western Himalayas, Eastern Himalayas

128. *Mnium marginatum* var. *riparium* (Mitt.) Husn.; Present status: **Synonym of** *Polla riparia* (Mitt.) Loeske

Distribution in India: Western Himalayas, Eastern Himalayas

129. *Mnium maximoviczii* Lindb.; Present status: **Synonym of** *Plagiomnium maximoviczii* (Lindb.) T.J. Kop.

Distribution in India: Western Himalayas, Eastern Himalayas

130. *Mnium medium* Bruch & Schimp.; Present status: **Synonym of** *Plagiomnium medium* (Bruch & Schimp.) T.J. Kop.

Distribution in India: Western Himalayas, Eastern Himalayas

131. *Mnium orthorrhynchum* Brid.; Present status: **Synonym of** *Polla orthorrhyncha* (Brid.) Brid. & Loeske

Distribution in India: Western Himalayas, Eastern Himalayas

132. *Mnium punctatum* Hedw.; Present status: **Synonym of** *Rhizomnium punctatum* (Hedw.) T.J. Kop.

Distribution in India: Western Himalayas, Eastern Himalayas

133. *Mnium punctatum* var. *horikawae* (Nog.) Nog.;Present status: **Synonym of** *Rhizomnium punctatum* var. *horikawae* (Nog.) Nog.

Distribution in India: Western Himalayas, Eastern Himalayas

134. *Mnium rostratum* Schrad.; Present status: **Synonym of** *Pohlia mauiensis* (Broth. & E.B. Bartram) W. Schultze-Motel

Distribution in India: Western Himalayas

135. *Mnium seligeri* Jur.; Present status: **Synonym of** *Plagiomnium elatum* (Bruch & Schimp.) T.J. Kop.

Distribution in India: Western Himalayas

136. *Mnium stellare* Reichard & Hedw.; **Present status: Valid name**

Distribution in India: Western Himalayas

137. *Mnium striatulum* Mitt.; Present status: **Synonym of** *Rhizomnium striatulum* (Mitt.) T.J. Kop.

Distribution in India: Western Himalayas, Eastern Himalayas

138. *Mnium succulentum* Mitt.; Present status: **Synonym of** *Plagiomnium succulentum* (Mitt.) T.J. Kop.

Distribution in India: Western Himalayas, Eastern Himalayas and South India

139. *Mnium thomsonii* Schimp.; **Present status: Valid name**

Distribution in India: Western Himalayas, Eastern Himalayas

140. *Mnium undulatum* (Lindb.) Müll. Hal.; **Present status: Valid name**

Distribution in India: Eastern Himalayas

141. *Mnium vesicatum* Besch.; Present status: **Synonym of** *Plagiomnium vesicatum* (Besch.) T.J. Kop.

Distribution in India: Eastern Himalayas

142. *Mnium vesicatum* var. *latedecurrens* Kabiersch; **Synonym of** *Plagiomnium vesicatum* (Besch.) T.J. Kop.

Distribution in India: Eastern Himalayas

15. *Orthomniopsis* Broth.

143. *Orthomniopsis dilatata* (Wilson ex Mitt.) Nog.; **Present status: Valid name**

Distribution in India: Eastern Himalayas

16. *Orthomnion* (Mitt.) Broth.

144. *O. bryoides* (Griff.) Nork.; **Present status: Valid name**

17. *Pohlia* Hedw.

145. *Pohlia ampullacea* Hampe; **Present status: Doubtful name**

Distribution in india: Eastern Himalayas (Endemic to India)

146. *Pohlia camptotrachela* (Renauld & Cardot) Broth.; **Present status: Valid name**

Distribution in India: Western Himalayas, Eastern Himalayas and South India

147. *Pohlia cruda* (Hedw.) Lindb.; **Present status: Valid name**

Distribution in India: Western Himalayas, Eastern Himalayas

148. *Pohlia crudoides* (Sull. & Lesq.) Broth. ; **Present status: Valid name**

Distribution in India: Western Himalayas

149. *Pohlia elongata* Hedw.; **Present status: Valid name**

Distribution in India: Western Himalayas, Eastern Himalayas and South India

150. *Pohlia flexuosa* Harv. ; **Present status: Valid name**

Distribution in India: Western Himalayas, Eastern Himalayas, Gangetic plains and South India

151. *Pohlia flexuosa* var. *propagulifera* (Renauld & Cardot) Gangulee; **Present status: Valid name**

Distribution in india: Eastern Himalayas

152. *Pohlia gedeana* (Bosch & Sande Lac.) Gangulee; **Present status: Valid name**

Distribution in india: India Orientalis

153. *Pohlia himalayana* (Mitt.) Broth.; **Present status: Valid name**

Distribution in india: Eastern Himalayas (Endemic to India)

154. *Pohlia longicollis* (Hedw.) Lindb.; **Present status: Valid name**

Distribution in India: Western Himalayas, Eastern Himalayas

155. *Pohlia minor* Schleich. & Schwägr. ; **Present status: Valid name**

Distribution in India: Western Himalayas, Eastern Himalayas

156. *Pohlia minor* subsp. *acuminata* (Hoppe & Hornsch.) Wijk & Margad.; Present status: **Synonym of** *Pohlia elongata* Hedw.

Distribution in India: Western Himalayas, Eastern Himalayas and South India

157. *Pohlia rigescens* (Mitt.) Broth.; **Present status: Valid name**

Distribution in india: Eastern Himalayas (Endemic to India)

18. *Plagiomnium* T. J. Kop.

158. *Plagiomnium rostratum* (Schrad.) T. J. Kop.; **Present status: Valid name**

Distribution in India: Western Himalayas

Family: Aulacomniaceae Schimp.

19. *Aulocomnium* Schwägr.

159. *Aulocomnium androgynum* (Hedw.) Schwägr.; **Present status: Valid name**

Distribution in India: Western Himalayas

160. *Aulocomnium palustre* (Hedw.) Schwägr.; **Present status: Valid name**

Distribution in India: Western Himalayas

161. *Aulocomnium turgidum* (wahlenb.) Schwägr.; **Present status: Valid name**

Distribution in India: Western Himalayas

Family : Bartramiaceae Schwägr.

20. *Anacolia* Schimp.

162. *Anacolia menziesii* (Turner) Paris; **Present status: Valid name**

Distribution in India: Eastern Himalayas

163. *Anacolia sinensis* Broth.; Present status: **Synonym of** *Flowersia sinensis* (Broth.) D.G. Griffin & W.R. Buck.

Distribution in India: Eastern Himalayas

21. *Bartramia* Hedw.

164. *Bartramia gathica* **Cardot & P. de la Varde; Synonym of** *Bartramia brevifolia* subsp. *brevifolia*

Distribution in India: South India

165. *Bartramia halleriana* Hedw. ; **Present status: Valid name**

Distribution in India: Western Himalayas, Eastern Himalayas

166. *Bartramia ithyphylla* Brid. ; **Present status: Valid name**

Distribution in India: Western Himalayas, Eastern Himalayas

167. *Bartramia leptodonta* Wilson; Present status: **Synonym of** *Bartramia brevifolia* subsp. *brevifolia*

Distribution in India: Western Himalayas, Eastern Himalayas and South India

168. *Bartramia madurensis* **Dixon & P. de la Varde; Present status: Doubtful name**

Distribution in India: South India

169. *Bartramia pomiformis* Hedw.; **Present status: Valid name**

Distribution in India: Western Himalayas, Eastern Himalayas

170. *Bartramia pomiformis* var. *elongata* Turner ; Present status: **Synonym of** *Bartramia pomiformis* Hedw.

Distribution in India: Western Himalayas, Eastern Himalayas

171. *Bartramia subpellucida* Mitt.; Present status: **Synonym of** *Bartramia brevifolia* subsp. *brevifolia*

Distribution in India: Western Himalayas, Eastern Himalayas

172. *Bartramia subulata* Bruch & Schimp.; **Present status: Valid name**

Distribution in India: Western Himalayas, Eastern Himalayas and South India

22. *Bartramidula* Bruch & Schimp.

173. *Bartramidula bartramioides* (Griff.) Wijk. & Margad ; **Present status: Valid name**

Distribution in India: Western Himalayas, Eastern Himalayas and South India

174. *Bartramidula dispersa* Cardot & P. de la Varde; Present status: **Synonym of** *Philonotis dispersa* (Cardot & P. de la Varde) D.G. Griffin & W.R. Buck

Distribution in India: South India

175. *Bartramidula griffithiana* var. *sikkimensis* Kabiersch ; Present status: **Synonym of** *Philonotis sikkimensis* (Kabiersch) T.J. Kop.

Distribution in India: Western Himalayas, Eastern Himalayas and South India

176. *Bartramidula roylei* (Hook. f.) Bruch & Schimp.; Present status: **Synonym of** *Glyphocarpus roylei* (Hook. f.) A. Jaeger

Distribution in India: Western Himalayas, Central India, Gangetic plains and South India

23. *Breutelia* Bruch. & Schimp.

177. *Breutelia arundinifolia* (Duby) M. Fleisch.; **Present status: Valid name**

Distribution in India: Indian Orientalis

178. *Breutelia dicranacea* (Müll. Hal.) Mitt.; **Present status: Valid name**

Distribution in India: Western Himalayas, Eastern Himalayas and South India

179. *Breutelia sclerodictya* Cardot; Present status: **Synonym of** *Breutelia microdonta* (Mitt.) Broth.

Distribution in India: South India

24. *Conostomum* Sw. *in* Web. & Mohr

180. *Conostomum tetragonum* (Hedw.) Lindb.; **Present status: Valid name**

Distribution in India: Western Himalayas

25. *Fleischerobryum* Loesk.

181. *Fleischerobryum longicolle* (Hampe) Loesk.; **Present status: Valid name**

Distribution in India: Western Himalayas, Eastern Himalayas

182. *Fleischerobryum macrophyllum* Broth.; **Present status: Valid name**

Distribution in India: Eastern Himalayas

26. ***Philonotis*** Brid.

183. *Philonotis alpicola* Jur.; **Present status: Doubtful name**

Distribution in India: South India

184. *Philonotis angusta* Mitt.; Present status: **Synonym of** *Bartramia angusta* (Mitt.) Müll. Hal.

Distribution in India: Western Himalayas, Eastern Himalayas

185. *Philonotis anisoclada* Cardot & P. de la Varde; Present status: **Synonym of** *Philonotis falcata* (Hook.) Mitt.

Distribution in India: South India

186. *Philonotis calcarea* (Bruch & Schimp.) Schimp.; Present status: **Synonym of** *Philonotis fontana* (Hedw.) Brid.

Distribution in India: Western Himalayas

187. *Philonotis falcata* (Hook.) Mitt. (Asthana & Sahu, 2013); **Present status: Valid name**

Distribution in India: Western Himalayas, Eastern Himalayas, South India and Gangetic plains

188. *Philonotis fontana* (Hedw.) Brid.; **Present status: Valid name**

Distribution in India: Western Himalayas, Eastern Himalayas and South India

189. *Philonotis fontana* var. *pumilla* (Turner) Brid.; **Present status: Valid name**

Distribution in India: Western Himalayas

190. *Philonotis glomerata* Mitt.; Present status: **Synonym of** *Mielichhoferia glomerata* (Mitt.) Ochi

Distribution in India: Eastern Himalayas and South India

191. *Philonotis hastata* (Duby) Wijk & Margard ; **Present status: Valid name**

Distribution in India: Western Himalayas, Eastern Himalayas, South India and Gangetic plains

192. *Philonotis heterophylla* Mitt.; Present status: **Synonym of** *Philonotis hastata* (Duby) Wijk & Margad.

Distribution in India: Western Himalayas and South India

193. *Philonotis lancifolia* Mitt.; **Present status: Valid name**

Distribution in India: Western Himalayas

194. *Philonotis leptocarpa* Mitt.; Present status: **Synonym of** *Bartramia leptocarpa* (Mitt.) Müll. Hal.

Distribution in India: Western Himalayas, Eastern Himalayas

195. *Philonotis mollis* (Dozy & Molk.) Mitt.; **Present status: Valid name**

Distribution in India: Eastern Himalayas, South India and Andaman Islands

196. *Philonotis nitida* Mitt.; Present status: **Synonym of** *Philonotis turneriana* (Schwägr.) Mitt.

Distribution in India: Western Himalayas, Eastern Himalayas

197. *Philonotis revoluta* Bosch & Sande Lac.; Present status: **Synonym of** *Philonotis thwaitesii* Mitt.

Distribution in India: Eastern Himalayas

198. *Philonotis secunda* (Dozy & Molk.) Bosch & Sande Lac.; **Present status: Valid name**

Distribution in India: South India

199. *Philonotis seriata* Mitt.; Present status: **Synonym of** *Philonotis fontana* (Hedw.) Brid.

Distribution in India: Western Himalayas

200. *Philonotis subrigida* Cardot & P. de la Varde; **Present status: Doubtful name**

Distribution in India: South India

201. *Philonotis tomentella* Molendo; **Present status: Synonym of** *Philonotis fontana* (Hedw.) Brid

Distribution in India: South India

202. *Philonotis trachyphylla* Dixon & Badhw.; **Present status:** Doubtful name

Distribution in India: Western Himalayas

203. *Philonotis turneriana* (Schwägr.) Mitt.; **Present status: Valid name**

Distribution in India: Western Himalayas, Eastern Himalayas

204. *Philonotis thwaitesii* Mitt.; **Present status: Valid name**

Distribution in India: Western Himalayas, Eastern Himalayas, South India and Central India

27. *Plagiopus* Brid.

205. *Plagiopus oederi* (Brid.) Limpr.;Present status: **Synonym of** *Plagiopus oederianus* (Sw.) H.A. Crum & L.E. Anderson

Distribution in India: Western Himalayas

E. ORDER: **DICRANALES** H. PHILIB. & M. FLEISCH.

Family: Fissidentaceae Schimp.

28. *Fissidens* Hedw.

206. *Fissidens aberrans* Broth. & Dixon; Present status: **Synonym of** *Fissidens curvatus* Hornsch.

Distribution in India: South India

207. *Fissidens alanii* Gangulee; Present status: **Synonym of** of *Fissidens longisetus* Griff.

Distribution in India: Eastern Himalayas (Endemic to India)

208. *Fissidens angustiusculus* Dixon & P. de la Varde; Present status: **Doubtful name**

Distribution in India: South India

209. *Fissidens anomalus* Mont.; **Present status: Valid name**

Distribution in India: Eastern Himalayas, Western Himalayas and South India

210. *Fissidens areolatus* Griff.; Present status: **Synonym of** *Fissidens polypodioides* Hedw.

Distribution in India: Eastern Himalayas

211. *Fissidens arnigadhensis* Broth. & S. S. Kumar ; Present status: **Doubtful name**

Distribution in India: Western Himalayas

212. *Fissidens asplenioides* Hedw.; **Present status: Valid name**

Distribution in India: South India

213. *Fissidens asperisetus* Sande Lac.; **Synonym of** *Fissidens hollianus var. asperisetus* (Sande Lac.) M. Fleisch.

Distribution in India: South India and Andaman Islands.

214. *Fissidens asperisetus* var. *andamanensis*; Present status: **Doubtful name**

Distribution in India: South India and Andaman Islands.

215. *Fissidens arunii* Srivastava & Norkett ; Present status: **Doubtful name**

Distribution in India: Western Himalayas and Central India

216. *Fissidens biformis* Mitt.; Present status: **Doubtful name**

Distribution in India: South India

217. *Fissidens bilaspurense* Gangulee; Present status: **Doubtful name**

Distribution in India: Central India, Gangetic plains (Endemic to India)

218. *Fissidens bryoides* Hedw.; **Present status: Valid name**

Distribution in India: Western Himalayas, Eastern Himalayas, South India, Gangetic plains

219. *Fissidens ceylonensis* Dozy & Molk.; **Present status: Valid name**

Distribution in India: Western Himalayas, Eastern Himalayas, South India, Gangetic plains

220. *Fissidens crenulatus* Mitt.; **Present status: Valid name**

Distribution in India: Central India, Gangetic plains

221. *Fissidens crenulatus* var. *titalyanus* (Müll. Hal.) Gangulee; **Present status: Valid name**

Distribution in India: Central India, Gangetic plains

222. *Fissidens cristatus* Wilson & Mitt.; Present status: **Synonym of** *Fissidens dubius* P. Beauv.

Distribution in India: Western Himalayas, Eastern Himalayas, South India

223. *Fissidens curvatoinvolutus* Dixon ; Present status: **Synonym of** *Fissidens involutus* subsp. *curvatoinvolutus* (Dixon) Gangulee

Distribution in India: Western Himalayas, Eastern Himalayas, Central India and Gangetic Plains

224. *Fissidens curvatoxiphioides* Dixon & P. de la Varde ; Present status: **Synonym of** *Fissidens beckettii* Mitt.*f*

Distribution in India: Western Himalayas and South India

225. *Fissidens diversifolius* Mitt.; **Present status: Valid name**

Distribution in India: Western Himalayas, Eastern Himalayas, South India, Central India and Gangetic Plains

226. *Fissidens elongatus* Mitt.; Present status: **Doubtful name**

Distribution in India: Eastern Himalayas (Endemic to India)

227. *Fissidens firmus* Mitt.; **Present status: Valid name**

Distribution in India: South India

228. *Fissidens ganguleei* Nork.; Present status: **Doubtful name**

Distribution in India: Eastern Himalayas and South India

229. *Fissidens geminiflorus* Dozy & Molk.; **Present status: Valid name**

Distribution in India: Central India

230. *Fissidens geppii* Fleisch.; Present status: **Doubtful name**

Distribution in India: Western Himalayas

231. *Fissidens grandifrons* Brid.; **Present status: Valid name**

Distribution in India: Western Himalayas and South India

232. *Fissidens griffithii* Gangulee; Present status: **Doubtful name**

Distribution in India: South India

233. *Fissidens hyalinus* Hook. & Wilson ; **Present status: Valid name**

Distribution in India: Western Himalayas

234. *Fissidens intromarginatulus* E.B. Bartram ; Present status: **Synonym of** *Fissidens ceylonensis* Dozy & Molk.

Distribution in India: Western Himalayas, Eastern Himalayas, South India and Central India

235. *Fissidens involutus* Wilson & Mitt.; **Present status: Valid name**

Distribution in India: Western Himalayas, Eastern Himalayas and Central India

236. *Fissidens javanicus* Dozy & Molk.; **Present status: Valid name**

Distribution in India: Western Himalayas, Eastern Himalayas and Andaman Islands

237. *Fissidens jungermannioides* Griff.; Present status: **Doubtful name**

Distribution in India: Eastern Himalayas

238. *Fissidens kalimpongensis* Gangulee ; Present status: **Doubtful name**

Distribution in India: Western Himalayas, Eastern Himalayas and South India

239. *Fissidens kurzii* Müll. Hal.; Present status: **Doubtful name**

Distribution in India: Gangetic plains

240. *Fissidens laxitextus* Broth. *ex* Gangulee; Present status: **Doubtful name**

Distribution in India: Eastern Himalayas

241. *Fissidens leptopelma* Dixon; Present status: **Synonym of** *Fissidens subangustus* M. Fleisch.

Distribution in India: Eastern Himalayas

242. *Fissidens longisetus* Griff.; **Present status: Valid name**

Distribution in India: Eastern Himalyas

243. *Fissidens macrosporoides* Dixon & P. de la Varde ; Present status: **Doubtful name**

Distribution in India: Western Himalayas, Central India and South India

244. *Fissidens minutus* Thwaites & Mitt.; Present status: **Synonym of** *Fissidens pallidinervis* Mitt.

Distribution in India: South India

245. *Fissidens mittenii* Paris; Present status: **Synonym of** *Fissidens laxus* Sull. & Lesq.

Distribution in India: South India

246. *Fissidens mussureinsis* C. Muell. in Bruhl. ; **Present status: Valid name**

Distribution in India: Western Himalayas

247. *Fissidens nobilis* Griff.; **Present status: Valid name**

Distribution in India: Western Himalayas, Eastern Himalayas

248. *Fissidens nymanii* Fleisch.; **Present status: Valid name**

Distribution in India: Western Himalayas, South India

249. *Fissidens obscurus* Mitt. ; **Present status: Valid name**

Distribution in India: Western Himalayas, Eastern Himalayas and South India

250. *Fissidens orishae* Gangulee ; **Present status: Valid name**

Distribution in India: Central India and Gangaetic plains

251. *Fissidens perplexans* Dixon ; Present status: **Doubtful name**

Distribution in India: Western Himalayas

252. *Fissidens perumalensis* Dixon & P. de la Varde; Present status: **Doubtful name**

Distribution in India: South India

253. *Fissidens pokhrensis* Nork. & S.S. Kumar ; Present status: **Doubtful name**

Distribution in India: Western Himalayas

254. *Fissidens polysetulus* Müll. Hal.; **Present status: Doubtful name**

Distribution in India: Eastern Himalayas

255. *Fissidens pulchellus* Mitt.; **Present status: Doubtful name**

Distribution in India: Eastern Himalayas

256. *Fissidens rambii* Gangulee; **Present status: Doubtful name**

Distribution in India: Eastern Himalayas

257. *Fissidens raiatensis* E.B. Bartram; **Present status: Doubtful name**

Distribution in India: Central India, South India

258.*Fissidens ranchiensis* Gangulee; **Present status: Doubtful name**

Distribution in India: Central India, South India

259. *Fissidens ranuii* Gangulee; **Present status: Valid name**

Distribution in India: Central India (Endemic to India)

260. *Fissidens rigidiusculus* Broth.; Present status: **Synonym of** *Fissidens geppii* M. Fleisch.

Distribution in India: Eastern Himalayas (Endemic to India)

261. *Fissidens robinsonii* Broth.; **Present status: Valid name**

Distribution in India: Andaman and Nicobar Islands

262. *Fissidens sedgwickii* Broth. & Dixon; **Present status: Doubtful name**

Distribution in India: South India

263. *Fissidens semperfalcatus* Dixon; **Present status: Valid name**

Distribution in India: Western Himalayas, Eastern Himalayas, Gangetic plains and Andaman Islands

264. *Fissidens serratus* Müll. Hal.; **Present status: Valid name**

Distribution in India: South India

265. *Fissidens splachnobryoides* Broth.; Present status: **Synonym of** *Fissidens flaccidus* Mitt.

Distribution in India: Western Himalayas, Eastern Himalayas, South India, Central India and Gangetic plains

266. *Fissidens subbryoides* Gangulee; **Present status: Valid name**

Distribution in India: Eastern Himalayas and Andaman Islands

267. *Fissidens subbrachyneuron* Thér. & P. de la Varde; **Present status: Doubtful name**

Distribution in India: South India

268. *Fissidens subfirmus* Dixon; **Present status: Doubtful name**

Distribution in India: South India

269. *Fissidens subpalmatus* Müll. Hal.; **Present status: Valid name**

Distribution in India: Central India, South India and Gangetic plains

270. *Fissidens subpulchellus* Nork.; **Present status: Valid name**

Distribution in India: Eastern Himalayas (Endemic to India)

271. *Fissidens sylvaticus* Griff.; **Synonym of** *Fissidens crispulus* Brid.

Distribution in India: Eastern Himalayas (Endemic to India)

272. *Fissidens sylvaticus* var. *auriculatus* (Mull. Hal.) Gangulee; **Present status: Doubtful name**

Distribution in India: Eastern Himalayas (Endemic to India)

273. *Fissidens taxifolius* Hedw.; **Present status: Doubtful name**

Distribution in India: Western Himalayas, Eastern Himalayas, South India, Gangetic plains and Andaman Islands

274. *Fissidens virens* Thwaites & Mitt.; **Present status: Valid name**

Distribution in India: Western Himalayas and South India

275. *Fissidens. viridulus* (Sw.) Wahlenb. (Asthana & Sahu, 2013); Present status: **Synonym of** *Fissidens bryoides* Hedw.

Distribution in India: Western Himalayas

276. *Fissidens walkeri* Broth.; **Present status: Valid name**

Distribution in India: South India

277. *Fissidens xiphioides* M. Fleisch.; Present status: **Synonym of** *Fissidens zollingeri* Mont.

Distribution in India: Western Himalayas, South India, Gangetic plains, Central India and Andaman Islands

278. *Fissidens zollingeri* Mont.; **Present status: Valid name**

Distribution in India: South India and Andaman Islands

29. *Octogonella* Dixon

279. *Octogonella scabrifolia* Dixon ; Present status: **Synonym of** *Rhachithecium papillosum* (R.S. Williams) Wijk & Margad.

Distribution in India: Western Himalayas

Family : **Rhachitheciaceae** H. Rob.

30. *Rhachithecium* Jolis

280. *Rhachithecium papillosum* (R.S. Williams) Wijk & Margad.; **Present status: Valid name**

Distribution in India: Eastern Himalayas and South India

281. *Rhachithecium perpusillum* (Thwaites & Mitt.) Broth. .; **Present status: Valid name**

Distribution in India: Eastern Himalayas and South India

Family: Rhabdoweisiaceae Limpr.

31. *Amphidium* Schimp.

282. *Amphidium lapponicum* (Hedw.) Schimp.; **Present status: Valid name**

Distribution in India: Western Himalayas

32. *Cynodontium* Schimp.

283. *Cynodontium gracilescens* (F. Weber & D. Mohr) Schimp.; **Present status: Valid name**

Distribution in India: Eastern Himalayas

33. *Oreoweisia* (Bruch & Schimp.) De Not.

284. *Oreoweisia brevidens* Herzog.; **Present status: Valid name**

Distribution in India: Eastern Himalayas (Endemic to India)

285. *Oreoweisia laxifolia* (Hook. f.) Kindb.; **Present status: Valid name**

Distribution in India: Eastern Himalayas, Western Himalayas and South India

34. *Rhabdoweisia* Bruch & Schimp.

286. *Rhabdoweisia crenulata* (Mitt.) H. Jameson ; **Present status: Valid name**

Distribution in India: Western Himalayas, Eastern Himalayas

287. *Rhabdoweisia crispata* (With.) Lindb.; **Present status: Valid name**

Distribution in India: Eastern Himalayas

Family: Dicranaceae Schimp.

35. *Atractylocarpus* Mitt.

288. *Atractylocarpus erectifolius* Mitt. ex Dixon; **Present status: Valid name**

Distribution in India: Eastern Himalayas (Endemic to India)

289. *Atractylocarpus sinensis* (Broth.) Herz.; **Present status: Valid name**

Distribution in India: Eastern Himalayas

36. *Aongstroemia* Bruch & Schimp.

290. *Aongstroemia orientalis* Mitt. ; **Present status: Valid name**

Distribution in India: Western Himalayas, Eastern Himalayas

37. *Brothera* Müll. Hal.

291. *Brothera capillifolia* Dixon; Present status: **Synonym of** *Dicranodontium asperulum* (Mitt.) Broth

Distribution in India: Eastern Himalayas

292. *Brothera himalayana* Broth. Endemic to India ; Present status: **Synonym of** *Atractylocarpus himalayanus* (Broth.) J.-P. Frahm

Distribution in India: Western and Eastern Himalayas (Endemic to India)

293. *Brothera leana* (Sull.) Müll. Hal.; **Present status: Valid name**

Distribution in India: Western Himalayas, Eastern Himalayas and South India

38. *Campylopodiella* Cardot

294. *Campylopodiella ditrichoides* Nog.; Present status: **Synonym of** *Campylopodiella himalayana* (Broth.) J.-P. Frahm

Distribution in India: Western Himalayas

295. *Campylopodiella tenella* Cardot.; **Present status: Valid name**

Distribution in India: Eastern Himalayas

39. *Campylopodium* (Müll. Hal.) Besch.

296. *Campylopodium griffithi* (Mitt.) Mitt. & Broth.; **Present status: Valid name**

Distribution in India: Eastern Himalayas, Western Himalayas and South India

297. *Campylopodium khasianum* (Griffiths) Paris; Present status: **Synonym of** *Microcampylopus khasianus* (Griffiths) Giese & J.-P. Frahm

Distribution in India: Eastern Himalyas and South India

298. *Campylopodium phascoides* (C. Muell.) Paris; Present status: **Doubtful name**

Distribution in India: South India

40. *Campylopus* Brid.

299. *Campylopus albescens* (Müll. Hal.) A. Jaeger; Present status: **Synonym of** *Campylopus fragilis* subsp. *zollingerianus* (Müll. Hal.) J.-P. Frahm

Distribution in India: South India

300. *Campylopus alpigena* Broth.; **Present status: Valid name**

Distribution in India: Eastern Himalayas

301. *Campylopus andreanus* Cardot & P. de la Varde; **Present status: Valid name**

Distribution in India: South India

302. *Campylopus aureus* Bosch & Sande Lac.; Present status: **Synonym of** *Campylopus schmidii* (Müll. Hal.) A. Jaeger

Distribution in India: South India

303. *Campylopus caudatus* (Müll. Hal.) Mont.; **Present status: Valid name**

Distribution in India: South India

304. *Campylopus comosus (Schwägr.)* Bosch & Sande Lac.; **Present status: Valid name**

Distribution in India: South India

305. *Campylopus durelii* Gangulee; **Present status: Valid name**

Distribution in India: South India

306. *Campylopus eberhardtii* Paris; **Present status: Valid name**

Distribution in India: South India

307. *Campylopus ericoides* **(Griff.) A. Jaeger; Present status: Valid name**

Distribution in India: Eastern Himalayas

308. *Campylopus erythrognaphalus* (Müll. Hal.) A. Jaeger; **Present status:** Synonym of *Campylopus ericoides* (Griff.) A. Jaeger

Distribution in India: South India

309. *Campylopus flexuosus* (Hedw.) Brid.; **Present status: Valid name**

Distribution in India: Eastern Himalayas

310. *Campylopus* fragilis (Brid.) Bruch & Schimp. ; **Present status: Valid name**

Distribution in India: Western Himalayas and South India

311. *Campylopus fragilis* var. *pyriformis* (Schultz.) Agst.; **Present status: Valid name**

Distribution in India: Western Himalayas and Eastern Himalayas

312. *Campylopus goughii* (Mitt.) A. Jaeger ; Present status: **Synonym of** *Campylopus fragilis* subsp. *zollingerianus* (Müll. Hal.) J.-P. Frahm

Distribution in India: Western Himalayas, Eastern Himalayas and South India

313. *Campylopus gracilis* (Mitt.) A. Jaeger.; **Present status: Valid name**

Distribution in India: Western Himalayas, Eastern Himalayas and South India

314. *Campylopus introflexus* (Hedw.) Brid.; **Present status: Valid name**

Distribution in India: Eastern Himalayas and South India

315. *Campylopus involutus* (Müll. Hal.) A. Jaeger; Present status: **Synonym of** *Thysanomitrion involutum* (Müll. Hal.) P. de la Varde

Distribution in India: Eastern Himalayas and South India

316. *Campylopus laetus* (Mitt.) A. Jaeger; Present status: **Synonym of** *Campylopus savannarum* (Müll. Hal.) Mitt.

Distribution in India: Eastern Himalayas and South India

317. *Campylopus latinervis* (Mitt.) A. Jaeger ; Present status: **Synonym of** *Campylopus gracilis* (Mitt.) A. Jaeger

Distribution in India: Eastern Himalayas and South India

318. *Campylopus milleri* Renauld & Cardot; **Present status: Valid name**

Distribution in India: Eastern Himalayas

319. *Campylopus nietneri (*Müll. Hal.) A. Jaeger; Present **status:** Synonym of *Campylopus fragilis* subsp. goughii (Mitt.) J.-P. Frahm

Distribution in India: South India

320. *Campylopus nilghiriensis* (Mitt.) A. Jaeger; **Present status:** Synonym of *Campylopus zollingerianus* (Müll. Hal.) Bosch & Sande Lac.

Distribution in India: South India

321. *Campylopus nodiflorus* (Müll. Hal.) A. Jaeger; **Present status:** Synonym of *Campylopus schmidii* (Müll. Hal.) A. Jaeger

Distribution in India: South India

322. *Campylopus pseudogracilis* Cardot & Dixon; **Present status:** Synonym of *Campylopus fragilis* subsp. *zollingerianus* (Müll. Hal.) J.-P. Frahm

Distribution in India: South India

323. *Campylopus recurvus* (Mitt.) A. Jaeger; **Present status: Valid name**

Distribution in India: South India

324. *Campylopus reduncus* (Reinw. & Hornsch.) Bosch & Sande Lac.; **Present status:** Synonym of *Campylopus comosus (Schwägr.)* Bosch & Sande Lac.

Distribution in India: South India

325. *Campylopus richardii* Brid.; **Present status: Valid name**

Distribution in India: Western Himalayas, Eastern Himalayas and South India

326. *Campylopus roinei* Cardot & P. de la Varde; **Present status:** Synonym of *Thysanomitrion involutum* (Müll. Hal.) P. de la Varde

Distribution in India: South India

327. *Campylopus schmidii* (Müll. Hal.) A. Jaeger; **Present status: Valid name**

Distribution in India: South India

328. *Campylopus sedgwickii* Dixon; **Present status:** Synonym of *Campylopus recurvus* (Mitt.) A. Jaeger

Distribution in India: South India

329. *Campylopus schwarzii* Schimp.; Present status: **Synonym of** *Campylopus gracilis* (Mitt.) A. Jaeger

Distribution in India: Eastern Himalayas

330. *Campylopus subfragilis* Renauld & Cardot; **Present status: Valid name**

Distribution in India: Eastern Himalayas and South India

331. *Campylopus subgracilis* Renauld & Cardot ex Gangulee; Present status: **Synonym of** *Campylopus zollingerianus* (Müll. Hal.) Bosch & Sande Lac

Distribution in India: Eastern Himalyas and South India

332. *Campylopus subluteus* (Mitt.) A. Jaeger; **Present status: Valid name**

Distribution in India: Eastern Himalyas

41. ***Dicranella*** (Müll. Hal.) Schimp.

333. *Dicranella divaricata* (Mitt.) A. Jaeger ; Present status: **Synonym of** *Aongstroemia divaricata* (Mitt.) Müll. Hal.

Distribution in India: Western Himalayas, Eastern Himalayas and South India

334. *Dicranella heteromalla* (Hedw.) Schimp.; **Present status: Valid name**

Distribution in India: Western Himalayas, Eastern Himalayas

335. *Dicranella leptoneura* Dixon; **Present status: Doubtful name**

Distribution in India: Eastern Himalayas (Endemic to India)

336. *Dicranella macrospora* Gangulee; **Present status: Doubtful name**

Distribution in India: Eastern Himalayas (Endemic to India)

337. *Dicranella psudosubulata* Mull. Hal. ex Gangulee; **Present status: Doubtful name**

Distribution in India: Eastern Himalayas (Endemic to India)

338. *Dicranella setifera* (Mitt.) A. Jaeger; **Present status: Valid name**

Distribution in India: Eastern Himalayas

339. *Dicranella spiralis* (Mitt.) A. Jaeger.; Present status: **Valid name**

Distribution in India: Eastern and Western Himalayas

42. ***Dicranodontium*** Bruch & Schimp.

340. *Dicranodontium asperulum* (Mitt.) Broth.; **Present status: Valid name**

Distribution in India: Eastern Himalayas

341. *Dicranodontium attenuatum* (Mitt.) Wilson & A. Jaeger ; **Present status: Valid name**

Distribution in India: Eastern and Western Himalayas

342. *Dicranodontium caespitosum* (Mitt.) Paris ; **Present status: Valid name**

Distribution in India: Eastern and Western Himalayas

343. *Dicranodontium capillifolium* (Dixon) Takaki; **Present status: Valid name**

Distribution in India: Eastern Himalayas (Endemic to India)

344. *Dicranodontium decipiens* (Mitt.) Mitt. ex Broth.; **Present status: Valid name**

Distribution in India: Eastern Himalayas

345. *Dicranodontium denudatum* (Brid.) E. Britton ; **Present status: Valid name**

Distribution in India: Eastern and Western Himalayas

346. *Dicranodontium didictyon* (Mitt.) A. Jaeger.; **Present status: Valid name**

Distribution in India: Eastern and Western Himalayas and South India

347. *Dicranodontium didymodon* (Griff.) Paris; **Present status: Valid name**

Distribution in India: Eastern Himalayas

348. *Dicranodontium dimorphum* Mitt.; **Present status: Synonym of** *Dicranodontium didictyon* (Mitt.) A. Jaeger

Distribution in India: Eastern Himalayas

349. *Dicranodontium fleischerianum* W. Schultze-Motel; **Present status: Synonym of** *Dicranodontium uncinatum* (Harv.) A. Jaeger

Distribution in India: Eastern Himalayas

350. *Dicranodontium perviride* Dixon & P. de la Varde; **Present status: Synonym of** *Dicranodontium denudatum* (Brid.) E. Britton

Distribution in India: South India

351. *Dicranodontium uncinatum* (Harv.) A. Jaeger; **Present status: Valid name**

Distribution in India: Eastern Himalayas

43. *Dicranoloma* (Renauld) Renauld

352. *Dicranoloma blumii* (Nees) Paris; **Present status: Valid name**

Distribution in India: Eastern Himalayas

353. *Dicranoloma gymnostomum* (Mitt.) Paris; **Present status: Synonym of** *Pseudochorisodontium gymnostomum* (Mitt.) C. Gao et al.

Distribution in India: Eastern Himalayas

354. *Dicranoloma subreflexifolium* (Müll. Hal.) Paris; **Present status: Valid name**

Distribution in India: Eastern Himalayas

44. *Dicranoweisia* Lindb. & Mild.

355. *Dicranoweisia alpina* (Mitt.) Paris; **Present status: Synonym of** *Hymenoloma alpinum* (Mitt.) Ochyra

Distribution in India: Eastern Himalayas

356. *Dicranoweisia brevifolia* Dixon *in* Herzog; **Present status: Synonym of** *Hymenoloma brevifolium* (Herzog) Ochyra

Distribution in India: India Orientalis

357. *Dicranoweisia cirrata* (Hedw.) Lindb.; **Present status: Valid name**

Distribution in India: Western Himalayas

358. *Dicranoweisia crispula* (Hedw.) Milde ; **Present status: Valid name**

Distribution in India: Western Himalayas

359. *Dicranoweisia indica* (Wilson) Paris ; **Present status: Valid name**

Distribution in India: Western Himalayas, Eastern Himalayas

45. *Dicranum* Hedw.

360. *Dicranum assamicum* Dixon; Present status: **Doubtful name**

Distribution in India: Eastern Himalayas (Endemic to India)

361. *Dicranum bonjeanii* De Not.; **Present status: Valid name**

Distribution in India: Western Himalayas

362. *Dicranum crispifolium* Müll. Hal.; Present status: **Doubtful name**

Distribution in India: Eastern Himalayas and South India

363. *Dicranum dilatinerve* Cardot & P. de la Varde; Present status: **Doubtful name**

Distribution in India: South India

364. *Dicranum gymnostomum* Mitt.; **Present status: Valid name**

Distribution in India: Eastern Himalayas

365. *Dicranum himalayanum* Mitt.; **Present status: Valid name**

Distribution in India: Western Himalayas and Eastern Himalayas

366. *Dicranum kashmirense* Broth.; Present status: **Doubtful name**

Distribution in India: Western Himalayas and Eastern Himalayas

367. *Dicranum lorifolium* Mitt.; Present status: **Doubtful name**

Distribution in India: Western Himalayas and Eastern Himalayas

368. *Dicranum scoparium* Hedw. (Saxena & Gangwar, 2005); **Present status: Valid name**

Distribution in India: Western Himalayas and Eastern Himalayas

369. *Dicranum spurium* Hedw.; **Present status: Valid name**

Distribution in India: Western Himalayas and Eastern Himalayas

370. *Dicranum undulatum* Schrad. & Brid. ; **Present status: Valid name**

Distribution in India: Western Himalayas and Eastern Himalayas

46. *Holomitrium* Brid.

371. *Holomitrium densifolium* (Wilson) Wijk & Margad.; **Present status: Valid name**

Distribution in India: Eastern Himalayas and South India

47. *Leucoloma* Brid.

372. *Leucoloma amoene-virens* Mitt.; Present status: **Synonym of** *Poecilophyllum amoene-virens* (Mitt.) Mitt.

Distribution in India: South India

373. *Leucoloma brevifolium* Dixon & P. de la Varde; Present status: **Synonym of** *Leucoloma amblyacron* Müll. Hal. ex Besch.

Distribution in India: South India

374. *Leucoloma malabarense* Besch. ex Renauld & Cardot; Present status: **Doubtful name**

Distribution in India: South India

375. *Leucoloma molle* (Müll. Hal.) Mitt.; **Present status: Valid name**

Distribution in India: South India

376. *Leucoloma nitens* (Thwaites & Mitt.) A. Jaeger; **Present status: Valid name**

Distribution in India: South India

377. *Leucoloma renauldii* Broth.; Present status: **Synonym of** *Leucoloma tenerum* Mitt.

Distribution in India: South India

378. *Leucoloma strictifolium* Dixon; Present status: **Synonym of** *Leucoloma tenerum* Mitt.

Distribution in India: South India

379. *Leucoloma tenerum* Mitt.; **Present status: Valid name**

Distribution in India: South India

380. *Leucoloma walkeri* Broth. ; Present status: **Synonym of** *Leucoloma taylorii* (Schwägr.) Mitt.

Distribution in India: South India

48. *Microcampylopus* (Müll. Hal.) Fleisch.

381. *Microcampylopus subanus* (C. Muell.) Fleisch.; **Present status: Valid name**

Distribution in India: Eastern Himalayas and South India

49. *Microdus* Schimp.

382. *Microdus assamicus* Dixon; Present status: **Synonym of** *Leptotrichella assamica* (Dixon) Ochyra

Distribution in India: Eastern Himalayas (Endemic to India)

383. *Microdus brasiliensis* (Duby) Thér.; Present status: **Synonym of** *Dicranella pomiformis* (Griff.) A. Jaeger

Distribution in India: Western Himalayas, Eastern Himalayas and South India

50. *Mitrobryum* H. Rob.

384. *Mitrobryum koelzi* H. Rob.; Present status: **Doubtful name**

Distribution in India: Western Himalayas

51. *Oncophorus* (Brid.) Brid.

385. *Oncophorus gracillimus* Dixon ; ;Present status: **Synonym of** *Oncophorus wahlenbergii* Brid.

Distribution in India: Western Himalayas

386. *Oncophorus virens* (Hedw.) Brid.; **Present status: Valid name**

Distribution in India: Western Himalayas, Eastern Himalayas

387. *Oncophorus wahlenbergii* Brid.; **Present status: Valid name**

Distribution in India: Western Himalayas, Eastern Himalayas

52. *Oreas* Brid.

388. *Oreas maritiana* (Hoppe & Hornsch.) Brid.; **Present status: Valid name**

Distribution in India: Western Himalayas, Eastern Himalayas and South India

53. *Orthodicranum* (Bruch & Schimp.) Loesk.

389. *Orthodicranum montanum* (Hedw.) Loesk.; **Present status: Valid name**

Distribution in India: Western Himalayas

54. *Paraleucobryum* (Limpr.) Loesk.

390. *Paraleucobryum enerve* (Thed.) Loesk.; **Present status: Valid name**

Distribution in India: Western Himalayas, Eastern Himalayas

391. *Paraleucobryum himalayanum* Dixon ; Present status: **Synonym of** *Campylopodiella himalayana* (Broth.) J.-P. Frahm

Distribution in India: Western Himalayas

55. ***Symblepharis*** Mont.

392. *Symblepharis reinwardtii* (Dozy & Molk.) Lac.; **Present status: Valid name**

Distribution in India: Eastern Himalayas

393. *Symblepharis vaginata* (Hook. & Harv.) Wijk & Margad.; **Present status: Valid name**

Distribution in India: Western Himalayas, Eastern Himalayas and South India

56. ***Trematodon*** Michx.

394. *Trematodon assamensis* Broth. In Roth.; **Present status: Doubtful**

Distribution in India: Eastern Himalayas

395. *Trematodon brevicalyx* Dixon; **Present status: Doubtful**

Distribution in India: Gangetic plains

396. *Trematodon capillifolius* Müll. Hal. & G. Roth ; Present status: **Doubtful name**

Distribution in India: Western Himalayas

397. *Trematodon conformis* Mitt.; **Present status: Valid name**

Distribution in India: Western Himalayas, Eastern Himalayas and South india

398. *Trematodon ceylonensis* Müll. Hal.; **Present status:** Synonym of *Trematodon longicollis* Michx.

Distribution in India: Eastern Himalayas, South India and Gangetic plains

399. *Trematodon hookeri* Müll. Hal.; **Present status: Doubtful**

Distribution in India: Eastern Himalayas

400. *Trematodon longicollis* Michx.; **Present status: Valid name**

Distribution in India: Western Himalayas, Eastern Himalayas, Gangetic plains and South india

401. *Trematodon megapophysatus* Müll. Hal.; **Present status: Doubtful**

Distribution in India: Eastern Himalayas

402. *Trematodon schmidii* Müll. Hal.; **Present status: Doubtful**

Distribution in India: South India

403. *Trematodon subulosus* Griff.; Present status: **Doubtful name**

Distribution in India: Western Himalayas, Eastern Himalayas and Punjab

Family: Ditrichaceae Limpr.

57. *Ceratodon* Brid.

404. *Ceratodon purpureus* (Hedw.) Brid.; **Present status: Valid name**

Distribution in India: Western Himalayas, Eastern Himalayas and South India

405. *Ceratodon stenocarpus* Bruch & Schimp.; Present status: **Synonym of** *Ceratodon purpureus* subsp. *stenocarpus* (Bruch & Schimp.) Dixon

Distribution in India: Western Himalayas, Eastern Himalayas and South India

58. *Distichium* Bruch & Schimp.

406. *Distichium capillaceum* (Hedw.) Bruch & Schimp.; **Present status: Valid name**

Distribution in India: Western Himalayas, Eastern Himalayas

407. *Distichium inclinatum* (Hedw.) Bruch & Schimp.; **Present status: Valid name**

Distribution in India: Western Himalayas, Eastern Himalayas

59. *Ditrichopsis* Broth.

408. *Ditrichopsis clausa* Broth.; Present status: **Doubtful name**

Distribution in India: Eastern Himalayas

60. *Ditrichum* Hampe

409. *Ditrichum amoenum* (Thwaites & Mitt.) Paris; **Present status: Valid name**

Distribution in India: South India

410. *Ditrichum apophysatum* Hampe ex Gangulee; Present status: **Doubtful name**

Distribution in India: Eastern Himalayas (Endemic to India)

411. *Ditrichum darjeelingense* Renauld & Cardot; Present status: **Synonym of** *Fissidens darjeelingensis* (Renauld & Cardot) P. Syd.

Distribution in India: Eastern Himalayas (Endemic to India)

412. *Ditrichum difficile* (Duby) M. Fleisch.; **Present status: Valid name**

Distribution in India: Western Himalayas and South India

413. *Ditrichum flexicaule* (Schwägr.) Hampe ; **Present status: Valid name**

Distribution in India: Western Himalayas and South India

414. *Ditrichum heteromallum* (Hedw.) Britt.; **Present status: Valid name**

Distribution in India: Western Himalayas and Eastern Himalayas

415. *Ditrichum homomallum* (Hedw.) Hampe ; Present status: **Synonym of** *Ditrichum heteromallum* (Hedw.) E. Britton

Distribution in India: Western Himalayas and Eastern Himalayas

416. *Ditrichum laxissimum* (Mitt.) Kuntze; **Present status: Valid name**

Distribution in India: Eastern Himalayas and South India (Endemic to India)

417.*Ditrichum longicrure* Renauld & Cardot; **Present status: Doubtful name**

Distribution in India: Eastern Himalayas

418. *Ditrichum pusillum* (Hedw.) Hampe ; **Present status: Valid name**

Distribution in India: Western Himalayas and Eastern Himalayas

419. *Ditrichum tortipes* (Mitt.) O. Kuntze ; **Present status: Valid name**

Distribution in India: Western Himalayas, Eastern Himalayas and South India

420. *Ditrichum tortipes* var. *strictum* Dixon & P. de la Varde; **Present status:** Doubtful name

Distribution in India: South India

421.*Ditrichum tortuloides* Grout ; Present status: **Synonym of** *Ditrichum ambiguum* Best

Distribution in India: Western Himalayas and Eastern Himalayas

61. *Garckea* C. Muell.

422. *Garckea abbreviata* Dixon & P. de la Varde; **Present status:** Doubtful name

Distribution in India: South India

62. *Saelania* Lindb.

423. *Saelania glaucescens* (Hedw.) Broth.; **Present status: Valid name**

Distribution in India: Western Himalayas

Family: Leucobryaceae Schimp.

63. *Leucobryum* Hampe

424. *Leucobryum aduncum* Dozy & Molk.; **Present status: Valid name**

Distribution in India: Eastern Himalayas, Western Himalayas and South India

426. *Leucobryum cucullifolium* Cardot ; Present status: **Synonym of** *Leucobryum humillimum* Cardot

Distribution in India: Eastern Himalayas, Western Himalayas

427. *Leucobryum humillimum* Cardot; **Present status: Valid name**

Distribution in India: South India and Punjab plains.

428. *Leucobryum imbricatum* Broth.; **Present status: Doubtful name**

Distribution in India: South India

429. *Leucobryum javense* (Brid.) Mitt.; **Present status: Valid name**

Distribution in India: Eastern Himalayas, Western Himalayas and South India

430. *Leucobryum juniperoideum* (Brid.) Müll. Hal.; **Present status: Valid name**

Distribution in India: Eastern Himalayas and South India

431. *Leucobryum mittenii* Besch.; Present status: **Synonym of** *Leucobryum humillimum* Cardot

Distribution in India: Eastern Himalayas

432. *Leucobryum neilgherrense* **Müll. Hal.; Present status: Valid name**

Distribution in India: Eastern Himalayas, Western Himalayas and South India

433. *Leucobryum sanctum* (Nees ex Schwägr.) Hampe ; **Present status: Valid name**

Distribution in India: Eastern Himalayas

434. *Leucobryum textorii* Besch.; Present status: **Synonym of** *Leucobryum neilgherrense* Müll. Hal.

Distribution in India: Eastern Himalayas and South India

64. *Ochrobryum* Mitt.

435. *Ochrobryum nepalense* Besch.; Present status: **Synonym of** *Ochrobryum kurzianum* Hampe

Distribution in India: Western Himalayas

436. *Ochrobryum propaguliferum* Dixon; Present status: **Synonym of** *Leucobryum humillimum* Cardot

Distribution in India: South India

65. *Pleuridiella* Robins.

437. *Pleuridiella colei* Robins. ; **Present status: Valid name**

Distribution in India: Eastern Himalayas

66. *Thiemea* C. Muell.

438. *Thiemea hampeana* Müll. Hal.; Present status: **Synonym of** *Wilsoniella hampeana* (Müll. Hal.) E.S. Salmon

Distribution in India: South India

Family: Erpodiaceae Broth.

67. *Aulacopilum* Wilson

439. *Aulacopilum tumidulum* Thwaites & Mitt.; **Present status: Valid name**

Distribution in India: South India

68. *Erpodium* (Brid.) Brid.

440. *Erpodium abbreviatum* (Mitt.) I. G. Stone ; **Present status: Valid name**

Distribution in India: Western Himalayas

441. *Erpodium luzonense* (E. B. Bartram) H. A. Crum ; **Present status: Valid name**

Distribution in India: Western Himalayas

442. *Erpodium mangiferae* Müll. Hal.; **Present status: Valid name**

Distribution in India: Western Himalayas, Eastern Himalayas, South India, Central India and Gangetic plains

69. *Solmsiella* C. Muell.

443. *Solmsiella biseriata* (Austin) Steere; **Present status: Valid name**

Distribution in India: Eastern Himalayas

444. *Solmsiella ceylonica* (Thwaites & Mitt.) Müll. Hal.; **Present status:** Synonym of *Solmsiella biseriata* (Austin) Steere

Distribution in India: South India

F. ORDER: DIPHYSCIALES M. FLEISCH.

Family: Diphysciaceae M. Fleisch.

70. *Diphyscium* D. Mohr

445. *Diphyscium involutum* Mitt.; **Present status: Synonym of** *Diphyscium mucronifolium* Mitt.

Distribution in India: Eastern Himalayas and South India

G. ORDER: ENCALYPTALES DIXON

Family: Encalyptaceae Schimp.

71. *Encalypta* Hedw.

446. *Encalypta alpina* G. L. Sm.; **Present status: Valid name**

Distribution in India: Western Himalayas

447. *Encalypta ciliata* Hedw.; **Present status: Valid name**

Distribution in India: Western Himalayas

448. *Encalypta rhabdocarpa* Schwägr.; Present status: **Synonym of** *Encalypta rhaptocarpa* Schwägr.

Distribution in India: Western Himalayas

449. *Encalypta streptocarpa* Hedw.; **Present status: Valid name**

Distribution in India: Western Himalayas

450. *Encalypta tibetana* Mitt.; **Present status:** Doubtful name

Distribution in India: Western Himalayas

451. *Encalypta vulgaris* Hedw. ; **Present status: Valid name**

Distribution in India: Western Himalayas

H. ORDER: **FUNARIALES** M. FLEISCH.

Family: Funariaceae Schwägr.

72. *Entosthodon* Schwägr.

451. *Entosthodon buseanus* Dozy & Molk.; **Present status: Valid name**

Distribution in India: South India

452. *Entosthodon diversinervis* Müll. Hal.; Present status: **Synonym of** *Funaria diversinervis* (Müll. Hal.) Broth

Distribution in India: South India

453. *Entosthodon pilifer* Mitt.; Present status: **Synonym of** *Funaria pilifera* (Mitt.) Broth.

Distribution in India: Western Himalayas

454. *Entosthodon perrottetii* Müll. Hal.; **Present status: Doubtful name**

Distribution in India: South India

455. *Entosthodon physcomitrioides* (Mont.) Mitt.; **Present status: Valid name**

Distribution in India: South India

456. *Entosthodon planifolius* Thwaites & Mitt.; Present status: **Synonym of** *Funaria planifolia* (Thwaites & Mitt.) Broth.

Distribution in India: South India

456. *Entosthodon pulchellus* (H. Philib.) Brugués; **Present status: Valid name**

Distribution in India: South India

457. *Entosthodon submarginatus* Müll. Hal.; Present status: **Synonym of** *Funaria submarginata* (Müll. Hal.) Broth.

Distribution in India: South India

458. *Entosthodon wallichii* Mitt.; Present status: **Synonym of** *Entosthodon buseanus* Dozy & Molk.

Distribution in India: Western Himalayas and Eastern Himalayas

459. *Entosthodon wichurae* M. Fleisch.; **Present status: Valid name**

Distribution in India: Western Himalayas and Eastern Himalayas

73. *Funaria* Hedw.

460. *Funaria calcarea* Wahlenb.; Present status: **Synonym of** *Funaria muehlenbergii* Turner

Distribution in India: Western Himalayas

461. *Funaria calcarea* var. *mediterranea* (Lindb.) C.E.O. Jensen & Medelius ;Present status: **Synonym of** *Funaria pulchella* H. Philib.

Distribution in India: Western Himalayas

462. *Funaria capillipes* (Müll. Hal. & Broth.) Broth.; **Present status: Valid name**

Distribution in India: Western Himalayas

463. *Funaria connivens* Müll. Hal.; **Present status:** Synonym of *Funaria hygrometrica* var. *calvescens* (Schwägr.) Mont.

Distribution in India: South India

464. *Funaria eberhardtii (*Broth. & Paris) Broth.; **Present status: Valid name**

Distribution in India: South India

465. *Funaria excurrentinervis* Cardot & P. de la Varde; **Present status: Valid name**

Distribution in India: South India

466. *Funaria hygrometrica* Hedw.; **Present status: Valid name**

Distribution in India: Western Himalayas, Eastern Himalayas, South India, Central India and Gangetic plains

467. *Funaria hygrometrica* var. *calvescens* (Schwägr.) Mont.; **Present status: Valid name**

Distribution in India: Western Himalayas, Eastern Himalayas, South India, Central India and Gangetic plains

468. *Funaria koelzei* E.B. Bartram ; **Present status:** Doubtful name

Distribution in India: Western Himalayas

469. *Funaria orthocarpa* Mitt.; Present status: **Synonym of** *Entosthodon orthocarpus* (Mitt.) A. Jaeger

Distribution in India: Western Himalayas

470.*Funaria submarginata* (Müll. Hal.) Broth.; **Present status: Valid name**

Distribution in India: South India

471. *Funaria wijkii* R. S. Chopra ; **Present status:** Doubtful name

Distribution in India: Western Himalayas

74. *Physcomitrium* (Brid.) Brid.

472. *Physcomitrium coorgense* Broth.; Present status: **Synonym of** *Brachymenium coorgense* (Broth.) Paris

Distribution in India: Eastern Himalayas, South India, Gangetic plains and Central India

473. *Physcomitrium cyathicarpum* Mitt.; Present status: **Synonym of** *Physcomitrium immersum* Sull.

Distribution in India: Western Himalayas, Central India, Rajasthan and Gangetic plains

474. *Physcomitrium eurystomum* Sendtn.; **Present status: Valid name**

Distribution in India: Western Himalayas, Central India, Punjab, Rajasthan and Gangetic plains

475. *Physcomitrium indicum* (Dixon) Gangulee; **Present status: Valid name**

Distribution in India: Gangetic plains

476. *Physcomitrium insigne* Dixon & P. de la Varde; **Present status:** Doubtful name

Distribution in India: South India

477. *Physcomitrium japonicum* (Hedw.) Mitt.; **Present status: Valid name**

Distribution in India: Western Himalayas, Eastern Himalayas and Gangetic plains

478. *Physcomitrium pulchellum* (Griff.) Mitt.; **Present status: Valid name**

Distribution in India: Eastern Himalayas and Gangetic plains

479. *Physcomitrium repandum* (Griff.) Mitt., **Present status: Valid name**

Distribution in India: Eastern Himalayas and Punjab

I. ORDER: **GRIMMIALES** M. FLEISCH.

Family: Grimmiaceae Arn.

75. *Coscinodon* Spreng.

480. *Coscinodon cribrosus* (Hedw.) Spruce ; **Present status: Valid name**

Distribution in India: Western Himalayas, Eastern Himalayas

76. *Grimmia* Hedw.

481. *Grimmia anodon* Bruch & Schimp.; **Present status: Valid name**

Distribution in India: Western Himalayas

482. *Grimmia apiculata* Hornsch.; **Present status: Synonym of** *Grimmia fuscolutea* Hook.

Distribution in India: Western Himalayas

483. *Grimmia apophysata* Gangulee; **Present status: Synonym of** *Grimmia indica* (Dixon & P. de la Varde) Goffinet & Greven

Distribution in India: Eastern Himalayas (Endemic to India)

484. *Grimmia donniana* G.L. Sm. & Spruce ; **Present status: Valid name**

Distribution in India: Western Himalayas

485. *Grimmia elongata* Kaulf.; **Present status: Valid name**

Distribution in India: Eastern Himalayas

486. *Grimmia indica* (Dixon & P. de la Varde) Goffinet & Greven; **Present status: Valid name**

Distribution in India: Eastern Himalayas and South India

487. *Grimmia khasiana* Mitt.; **Present status: Valid name**

Distribution in India: Eastern Himalayas(Endemic to India)

488. *Grimmia laevigata* Mitt.; Present status: **Synonym of** *Racomitrium laevigatum* A. Jaeger

Distribution in India: Western Himalayas, Eastern Himalayas

489. *Grimmia macrotheca* Mitt.; **Present status: Valid name**

Distribution in India: Eastern Himalayas

490. *Grimmia ovalis* (Hedw.) Lindb.; **Present status: Valid name**

Distribution in India: Western Himalayas, Eastern Himalayas and South India

491. *Grimmia pilifera* P. Beauv.; **Present status: Valid name**

Distribution in India: Western Himalayas

492. *Grimmia pulvinata* (Hedw.) G. L. Sm.; **Present status: Valid name**

Distribution in India: Western Himalayas

493. *Grimmia redunca* Wilson & Mitt.; Present status: **Synonym of** *Grimmia elongata* Kaulf.

Distribution in India: Western Himalayas, Eastern Himalayas

494. *Grimmia somervellii* Dixon; Present status: **Synonym of** *Grimmia longirostris* Hook.

Distribution in India: Eastern Himalayas

495. *Grimmia trichophylla* Grev.; **Present status: Valid name**

Distribution in India: Eastern Himalayas

496. *Grimmia unicolour* Hook.; **Present status: Valid name**

Distribution in India: Western Himalayas

77. *Racomitrium* Brid.

497. *Racomitrium canescens* (Hedw.) Brid.; Present status: **Synonym of** *Niphotrichum canescens* (Hedw.) Bednarek-Ochyra & Ochyra

Distribution in India: Western Himalayas, Eastern Himalayas

498. *Racomitrium crispipilum* (Taylor) A. Jaeger ; **Present status: Valid name**

Distribution in India: Western Himalayas, Eastern Himalayas and South India

499. *Racomitrium fuscescens* Wilson ; Present status: **Synonym of** *Bucklandiella fuscescens* (Wilson) Bednarek-Ochyra & Ochyra

Distribution in India: Western Himalayas, Eastern Himalayas

500. *Racomitrium heterostichum* (Hedw.) Brid.; **Present status: Valid name**

Distribution in India: Eastern Himalayas

501. *Racomitrium himalayanum* (Mitt.) A. Jaeger.; **Present status: Valid name**

Distribution in India: Western Himalayas, Eastern Himalayas

502. *Racomitrium lonuginosum* (Hedw.) Brid.; **Present status: Valid name**

Distribution in India: Eastern Himalayas

503. *Racomitrium strictifolium* (Mitt.) Jaeger; **Present status: Valid name**

Distribution in India: Eastern Himalayas

504. *Racomitrium subsecundum* (Hook. et Grev. ex Harv.) Mitt.; **Present status: Valid name**

Distribution in India: Western Himalayas, Eastern Himalayas and South India

78. *Schistidium* Brid.

505. *Schistidium alpicola* (Hedw.) Limpr.; Present status: **Synonym of** *Schistidium rivulare* (Brid.) Podp.

Distribution in India: Western Himalayas

506. *Schistidium apocarpum* (Hedw.) Bruch & Schimp.; **Present status: Valid name**

Distribution in India: Western Himalayas, Eastern Himalayas

Family: Ptychomitriaceae Schimp.

79. *Ptychomitrium* Fuernr.

507. *Ptychomitrium indicum* (Schrad.) A. Jaeger; **Present status: Valid name**

Distribution in India: Eastern Himalyas

508. *Ptychomitrium fauriei* Besch.; **Present status: Valid name**

Distribution in India: Eastern Himalyas

509. *Ptychomitrium tortula* (Harv.) A. Jaeger.; **Present status: Valid name**

Distribution in India: Western Himalayas, Eastern Himalayas and South India

Family: Drummondiaceae (Vitt) Goffn&, comb. nov.

80. *Drummondia* Hook. in Drum.

510. *Drummondia duthiei* Mitt. & Müll. Hal. .; Present status: **Synonym of** *Drummondia sinensis* Müll. Hal.

Distribution in India: Western Himalayas

511. *Drummondia stricta* (Mitt.) Müll. Hal. .; **Present status: Valid name**

Distribution in India: Western Himalayas, Eastern Himalayas and Andaman Islands

512. *Drummondia thomsonii* Mitt. .; **Present status: Valid name**

Distribution in India: Western Himalayas

J. ORDER:ORTHOTRICHALES DIXON

Family: Orthotrichaceae Arn.

81. *Desmotheca* Lindb.

513. *Desmotheca apiculata* (Doz. et Molk.) Lindb. in Cardot; **Present status: Valid name**

Distribution in India: Andaman Island

82. *Groutiella* Steere

514. *Groutiella goniorrhyncha* (Dozy & Molk.) E.B. Bartram; **Present status: Valid name**

Distribution in India: Eastern Himalayas and Andaman Island

83. *Macrocoma* (Müll. Hal.) Grout

515. *Macrocoma orthotrichoides* (Raddi) Wijk & Margad ; **Present status: Valid name**

Distribution in India: Western Himalayas

84. *Macromitrium* Brid.

516. *Macromitrium calymperoideum* Mitt.; Present status: **Doubtful name**

Distribution in India: Eastern Himalayas and South India

517. *Macromitrium hamatum* E.B. Bartram; Present status: **Synonym of** *Macromitrium macrosporum* Broth.

Distribution in India: Eastern Himalayas (Endemic to India)

518. *Macromitrium hymenostomum* Mont.; **Present status: Valid name**

Distribution in India: Western Himalayas

519. *Macromitrium incrustatifolium* H. Rob; Present status: **Doubtful name**

Distribution in India: Eastern Himalayas

520. *Macromitrium moorcroftii* (Hook. & Grev.) Schwägr.; **Present status: Valid name**

Distribution in India: Eastern Himalayas, Eastern Himalayas and South India

521. *Macromitrium nepalense* (Hook. & Grev.) Schwägr.; **Present status: Valid name**

Distribution in India: Eastern Himalayas and South India

522. *Macromitrium perrottetii* Müll. Hal.; Present status: **Synonym of** *Macrocoma tenuis* subsp. *sullivantii* (Müll. Hal.) Vitt

Distribution in India: Eastern Himalayas and South India

523. *Macromitrium rigbyanum* Dixon [Singh et al., 2010]; Present status: **Doubtful name**

Distribution in India: Eastern Himalayas (Endemic to India)

524. *Macromitrium turgidum* Müll. Hal.; Present status: **Doubtful name**

Distribution in India: Eastern Himalayas (Endemic to India)

85. *Orthotrichum* Hedw.

525. *Orthotrichum affine* Schwägr.; Present status: **Synonym of** *Orthotrichum affine* Schrad. & Brid.

Distribution in India: Western Himalayas

526. *Orthotrichum affine* var. *fastigiatum* (Bruch & Brid.) Huebener ; Present status: **Synonym of** *Orthotrichum affine* Schrad. & Brid.

Distribution in India: Western Himalayas

527. *Orthotrichum alpestre* Hornsch. &. B. S. G. ; **Present status: Valid name**

Distribution in India: Western Himalayas

528. *Orthotrichum anomalum* Hedw.; **Present status: Valid name**

Distribution in India: Western Himalayas and South India

529. *Orthotrichum cupulatum* Hoffm. & Brid.; **Present status: Valid name**

Distribution in India: Western Himalayas

530. *Orthotrichum duthiei* Venturi ; Present status: **Synonym of** *Orthotrichum alpestre* Hornsch. & B.S.G.

Distribution in India: Western Himalayas

531. *Orthotrichum griffithi* Mitt. & Dixon ; Present status**: Doubtful name**

Distribution in India: Western Himalayas and Eastern Himalayas

532. *Orthotrichum* hookeri Wilson & Mitt.; **Present status: Valid name**

Distribution in India: Western Himalayas and Eastern Himalayas

533. *Orthotrichum macounii* Austin ; **Present status: Valid name**

Distribution in India: Western Himalayas and Eastern Himalayas

534. *Orthotrichum obtusifolium* Brid.; **Present status: Valid name**

Distribution in India: Western Himalayas

535. *Orthotrichum pumilum* Sw.; **Present status: Valid name**

Distribution in India: Western Himalayas

536. *Orthotrichum rupestre* Schleich. & Schwägr. ; **Present status: Valid name**

Distribution in India: Western Himalayas and Eastern Himalayas

537. *Orthotrichum sikkimense* Herzog; Present status: **Synonym of** *Orthotrichum hookeri* Wilson ex Mitt.

Distribution in India: Eastern Himalayas

538. *Orthotrichum speciosum* Nees ; **Present status: Valid name**

Distribution in India: Western Himalayas and Eastern Himalayas

539. *Orthotrichum striatum* Hedw.; Present status: **Synonym of** *Dorcadion striatum* (Hedw.) Lindb.

Distribution in India: Western Himalayas

540. *Orthotrichum sturmii* Hoppe & Hornsch.; Present status: **Synonym of** *Orthotrichum rupestre* subsp. *sturmii* (Hoppe & Hornsch.) Boulay

Distribution in India: Western Himalayas

541. *Orthotrichum urnigerum* Myrin ; **Present status: Valid name**

Distribution in India: Western Himalayas

542. *Orthotrichum venustum* Venturi ; Present status: **Synonym of** *Orthotrichum alpestre* Hornsch. & B.S.G.

Distribution in India: Western Himalayas

543. *Orthotrichum virens* Venturi ; Present status: **Synonym of** *Orthotrichum crenulatum* Mitt.

Distribution in India: Western Himalayas and South India

86. *Schlotheimia* Brid.

544. *Schlotheimia grevilleana* Mitt.; **Present status: Valid name**

Distribution in India: Eastern Himalayas and

K. ORDER: **HEDWIGIALES** OCHYRA

Family: Hedwigiaceae Schimp.

87. *Braunia* Bruch & Schimp.

545. *Braunia apiculata* Cardot.; **Present status: Valid name**

Distribution in India: South India

546. *Braunia attenuata* (Mitt.) A. Jaeger ; **Present status: Valid name**

Distribution in India: Western Himalayas

547. *Braunia secunda* (Hook.) Bruch & Schimp.; **Present status: Valid name**

Distribution in India: South India

88. *Hedwigia* P. Beauv.

548. *Hedwigia ciliata* (Hedw.) P. Beauv.; **Present status: Valid name**

Distribution in India: Western Himalayas

89. *Hedwigidium* Bruch & Schimp.

549. *Hedwigidium integrifolium* (P. Beauv.) Dixon; **Present status: Valid name**

Distribution in India: Eastern Himalayas and South India

90. ***Kashyapia*** R. S. Chopra

550. *Kashyapia ambiguua* R.S. Chopra ; **Present status:** Doubtful name

Distribution in India: Western Himalayas **and** Eastern Himalayas

L. ORDER: **HOOKERIALES** M. FLEISCH

Family: Daltoniaceae Schimp.

91. ***Actinodontium*** Schwaegr.

551. *Actinodontium rhaphidostegum* (Müll. Hal.) Bosch & Sande Lac.; **Present status: Valid name**

Distribution in India: Eastern Himalayas and South India

92. ***Daltonia*** Hook. & Taylor

552. *Daltonia apiculata* Mitt.; Present status: **Doubtful name**

Distribution in India: Eastern Himalayas

553. *Daltonia aristifolia* Renauld & Cardot; **Present status: Valid name**

Distribution in India: Eastern Himalayas

554. *Daltonia aristifolia* Ren. *et.* Card. ssp. *leptophylla* Broth. *ex* Fleisch.; Present status: **Doubtful name**

Distribution in India: Eastern Himalayas

555. *Daltonia brevipedunculata* Mitt.; Present status: **Doubtful name**

Distribution in India: South India

556. *Daltonia decolyi* Broth. *ex* Gangulee; Present status: **Doubtful name**

Distribution in India: Eastern Himalayas (Endemic to India)

557. *Daltonia flexifolia* Mitt.; Present status: **Synonym of** *Daltonia marginata* Griff.

Distribution in India: Eastern Himalayas

558. *Daltonia gemmipara* Dixon; Present status: **Doubtful name**

Distribution in India: Eastern Himalayas

559. *Daltonia himalayensis* Dixon & Herzog; Present status: **Doubtful name**

Distribution in India: Eastern Himalayas (Endemic to India)

560. *Daltonia marginata* Griff.; **Present status: Valid name**

Distribution in India: Eastern Himalayas

561. *Daltonia perlaxiretis* **Dixon;** Present status: **Synonym of** *Daltonia aristifolia* Renauld & Cardot

Distribution in India: Eastern Himalayas

562. *Daltonia semitorta* Mitt.; Present status: **Doubtful name**

Distribution in India: Eastern Himalayas (Endemic to India)

563. *Daltonia subangustifolia* Renauld & Cardot; Present status: **Doubtful name**

Distribution in India: Eastern Himalayas (Endemic to India)

564. *Daltonia subapiculata* Hampe *ex* Gangulee; Present status: **Doubtful name**

Distribution in India: Eastern Himalayas (Endemic to India)

Family: Hookeriaceae Schimp.

93. *Dendrocyathophorum* Dixon

565. *Dendrocyathophorum intermedium* Herzog; Present status: **Doubtful name**

Distribution in India: Eastern Himalayas

566. *Dendrocyathophorum paradoxum* (Broth.) Dixon; Present status: **Valid name**

Distribution in India: Eastern Himalayas

94. *Distichophyllum* Dozy & Molk.

567. *Distichophyllum decolyi* Gangulee; Present status: **Doubtful name**

Distribution in India: Eastern Himalayas (Endemic to India)

568. *Distichophyllum griffithii* (Mitt.) Paris; **Present status: Valid name**

Distribution in India: Eastern Himalayas

569. *Distichophyllum heterophyllum* (Wilson ex Mitt.) Paris; **Present status: Valid name**

Distribution in India: Eastern Himalayas

570. *Distichophyllum. humifusum* (Wilson & Mitt.) Paris ; **Present status: Valid name**

Distribution in India: Western Himalayas and Eastern Himalayas

571. *Distichophyllum madurense* Thér. & P. de la Varde; Present status: **Doubtful name**

Distribution in India: South India

572. *Distichophyllum mittenii* Bosch & Sande Lac.; **Present status: Valid name**

Distribution in India: Central India

573. *Distichophyllum montagneanum* (Müll. Hal.) Bosch & Sande Lac.; **Present status: Valid name**

Distribution in India: South India

574. *Distichophyllum obovatum* (Griff.) Paris; **Present status: Valid name**

Distribution in India: Eastern Himalayas

575. *Distichophyllum succulentum* (Mitt.) Broth.; **Present status: Valid name**

Distribution in India: South India

576. *Distichophyllum schmidtii* Broth.; Present status: **Doubtful name**

Distribution in India: Western Himalayas and South India

94. *Eriopus* C. Muell.

577. *Eriopus remotifolius* Müll. Hal.; **Present status: Valid name**

Distribution in India: Eastern Himalayas

95. *Hookeria* J.E. Sm.

578. *Hookeria acutifolia* Hook. & Grev.; **Present status: Valid name**

Distribution in India: Western Himalayas, Eatern Himalayas and South India

96. *Hookeriopsis* (Besch.) A. Jaeger

579. *Hookeriopsis secunda* (Griff.) Broth.; **Present status: Synonym of** *Thamniopsis secunda* (Griff.) W.R. Buck

Distribution in India: Eastern Himalayas

580. *Hookeriopsis utacamundiana (*Mont.) Broth.; **Present status: Valid name**

Distribution in India: Eastern Himalayas and South India

Family: Leucomiaceae Broth.

97. *Leucomium* Mitt.

581. *Leucomium aneurodictyon* (Müll. Hal.) A. Jaeger; **Present status: Valid name**

Distribution in India: South India

582. *Leucomium decolyi* Broth. ex Gangulee; **Present status: Synonym of** *Hygrohypnum choprae* Vohra

Distribution in India: South India

Family: Pilotrichaceae Kindb.

98. *Callicostella* (Müll. Hal.) Mitt.

583. *Callicostella papillata* (Mont.) Mitt.; **Present status: Valid name**

Distribution in India: Eatern Himalayas, South India and Andaman Islands

99. *Lepidopilidium* (Müll. Hal.) Broth.

584. *Lepidopilidium furcatum* (Thwaites & Mitt.) Broth. ; **Present status: Valid name**

Distribution in India: South India

M. ORDER:HYPNALES (M. FLEISCH.) W. R. BUCK & VITT

Family: Ambystegiaceae G. Roth.

100. *Amblystegium* Schimp.

585. *Amblystegium sparsile* (Mitt.) A. Jaeger**; Present status: Valid name**

Distribution in India: Eastern Himalayas

Family: Campyliaceae (Kanda) W. R. Buck, comb. nov. (Amblystegiaceae subfam. Campyliodeae Kanda, J. Sci. Hiroshima Univ., Ser. B. Div. 2, Bot. 15:250. 1975[1976]).

101. ***Anacamptodon*** Brid.

586. *Anacamptodon validinervis* Dixon & P. de la Varde; **Present status:** Doubtful

Distribution in India: South India

102. ***Calliergon*** (Sull.) Kindb.

587. *Calliergon nubigenum* (Mitt.) Broth. **Present status: Valid name**

Distribution in India: Eastern Himalayas

Family: Cryphaeaceae Schimp.

103. ***Dendropogonella*** Britt.

588. *Dendropogonella rufescens* (Schimp.) E. Britton; **Present status: Valid name**

Distribution in India: South India

104. ***Hydrocryphaea*** Dixon

589. *Hydrocryphaea wardii* Dixon; **Present status: Doubtful name**

Distribution in India: Eastern Himalayas

105. ***Pilotrichopsis*** Besch.

590. *Pilotrichopsis dentata* (Mitt.) Besch.; **Present status: Valid name**

Distribution in India: Eastern Himalayas

591. *Pilotrichopsis ferruginea* (Mitt.) Broth.; **Present status: Valid name**

Distribution in India: South India

106. ***Schoenobryum*** Doz. & Molk.

592. *Schoenobryum concavifolium* (Griff.) Gangulee ; **Present status: Valid name**

Distribution in India: Eastern Himalayas and South India

593. *Schoenobryum julaceum* Dozy & Molk.; **Present status: Synonym of** *Schoenobryum concavifolium* (Griff.) Gangulee

Distribution in India: Eastern Himalayas and South India

107. ***Sphaerotheciella*** Fleisch.

594. *Sphaerotheciella sphaerocarpa* (Hook.) M. Fleisch.; **Present status: Valid name**

Distribution in India: Eastern Himalayas

Family: Thuidiaceae Schimp.

108. ***Abietinella*** Müll. Hal.

595. *Abietinella abietina* (Hedw.) M. Fleisch ; **Present status: Valid name**

Distribution in India: Western Himalayas and Eastern Himalayas

596. *Abietinella brandissi* (Müll. Hal.) A. Jaeger.; **Present status: Valid name**

Distribution in India: Western Himalayas (Endemic to India)

109. ***Actinothuidium*** (Besch.) Broth.

597.*Actinothuidium hookeri* (Mitt.) Broth.; **Present status: Valid name**

Distribution in India: Eastern Himalayas

598. *Actinothudium sikkimense* Warnst.; **Present status: Synonym of** *Actinothuidium hookeri* (Mitt.) Broth.

Distribution in India: Eastern Himalayas (Endemic to India)

110. ***Claopodium*** (Lesq. & Jam.) Ren. & Cardot.

599. *Claopodium assurgens* (Sull. & Lesq.) Cardot.; **Present status: Valid name**

Distribution in India: Western Himalayas, Eastern Himalayas and South India

600. *Claopodium nervosum* (Harv.) M. M. Fleisch.; *Present status:* **Synonym of** *Claopodium prionophyllum* (Müll. Hal.) Broth.

Distribution in India: Western Himalayas, Eastern Himalayas and South India

601. *Claopodium pellucinerve* (Mitt.) Best ; **Present status: Valid name**

Distribution in India: Western Himalayas, Eastern Himalayas and South India

602. *Claopodium prionophyllum* (Müll. Hal.) Broth. [Asthana & Sahu, 2013]; **Present status: Valid name**

Distribution in India: Western Himalayas, Eastern Himalayas and South India

111. *Haplocladium* (Müll. Hal.) Müll. Hal.

603. *Haplocladium angustifolium* (Hampe & Müll. Hal.) Broth.; **Present status: Valid name**

Distribution in India: Western Himalayas, Eastern Himalayas

604. *Haplocladium himalayanum* E. B. Bartram ; **Present status:** Synonym of *H. schimperi* Thér

Distribution in India: Western Himalayas

605. *Haplocladium microphyllum* (Hedw.) Broth. [Asthana & Sahu, 2013]; **Present status: Valid name**

Distribution in India: Eastern and Western Himalayas

606. *Haplocladium microphyllum* subsp. *capillatum* (Mitt.) Reimers ; **Present status: Valid name**

Distribution in India: Eastern and Western Himalayas

607. *Haplocladium schimperi* Ther.; **Present status: Valid name**

Distribution in India: Eastern and Western Himalayas

608. *Haplocladium stratosum* (Mitt.) Dixon ; **Present status: Valid name**

Distribution in India: Eastern and Western Himalayas

609. *Haplocladium vestitum* Dixon & P. de la Varde; **Present status:** Synonym of *Haplocladium microphyllum* subsp. *virginianum* (Brid.) Reimers

Distribution in India: South India

112. *Myurella* Bruch & Schimp.

610. *Myurella julacea* (Schwägr.) Schimp.; **Present status: Valid name**

Distribution in India: Western Himalayas

113. *Pelekium* Mitt.

611. *Pelekium bifarium* (Bosch & Sande Lac.) M. Fleisch; **Present status: Valid name**

Distribution in India: Eastern Himalayas

612. *Pelekium velatum* Mitt.; **Present status: Valid name**

Distribution in India: Eastern Himalayas

114. *Thuidium* Bruch & Schimp.

613. *Thuidium assimile* (Mitt.) A. Jaeger ; **Present status: Valid name**

Distribution in India: Western Himalayas and Eastern Himalayas

614. *Thuidium brotheri* E.S. Salmon; *Present status*: **Synonym of** *Pelekium investe* (Mitt.) Touw

Distribution in India: Eastern Himalayas and Central India

615. *Thuidium contortulum* (Mitt.) A. Jaeger.; **Present status: Valid name**

Distribution in India: Western Himalayas, Eastern Himalayas and South India

616. *Thuidium cymbifolium* (Dozy & Molk.) Dozy & Molk.; **Present status: Valid name**

Distribution in India: Western Himalayas, Eastern Himalayas, South India and Central India

617. *Thuidium glaucinum* (Mitt.) Bosch & Lac.; **Present status: Valid name**

Distribution in India: Western Himalayas, Eastern Himalayas and South India

618. *Thuidium haplohymenium* (Harv. & Hook. f.) A. Jaeger ; **Present status: Valid name**

Distribution in India: Western Himalayas and Eastern Himalayas

619. *Thuidium investe* (Mitt.) A. Jaeger; **Present status: Valid name**

Distribution in India: Eastern Himalayas

620. *Thuidium kiasense* R.S. Williams; **Present status: Valid name**

Distribution in India: Eastern Himalayas

621. *Thuidium koelzii* H. Rob.; *Present status*: **Synonym of** *Pelekium fuscatum* (Besch.) Touw

Distribution in India: Western Himalayas, Eastern Himalayas and Central India

622. *Thuidium meyenianum* (Hampe) Dozy & Molk.; **Present status: Valid name**

Distribution in India: Western Himalayas, Eastern Himalayas and South India

623. *Thuidium minusculum* (Mitt.) A. Jaeger.; **Present status: Valid name**

Distribution in India: Western Himalayas and Eastern Himalayas

624. *Thuidium orientale* Mitt. & Dixon ; Present status: **Synonym of** *Thuidium pristocalyx* (Müll. Hal.) A. Jaeger.

Distribution in India: Western Himalayas and Eastern Himalayas

625. *Thuidium philibertii* Limpr. [Asthana & sahu, 2013]; *Present status:* Synonym of *Thuidium recognitum* subsp. *philibertii* (Limpr.) Dixon

Distribution in India: Western Himalayas

626. *Thuidium sparsifolium* (Mitt.) A. Jaeger.; **Present status: Valid name**

Distribution in India: Western Himalayas and Eastern Himalayas

627. *Thuidium squarrosulum* Renauld & Cardot ; Present status: **Synonym of** *Pelekium minusculum* (Mitt.) Touw

Distribution in India: Western Himalayas, Eastern Himalayas and South India

628. *Thuidium stevensii* Renauld & Cardot; Present status: **Synonym of** *Thuidium glaucinum* (Mitt.) Bosch & Sande Lac.

Distribution in India: Eastern Himalayas

629. *Thuidium subpellucens* Dixon; Present status: **Synonym of** *Palamocladium wilkesianum* (Sull.) Müll. Hal.

Distribution in India: Eastern Himalayas (Endemic to India)

630. *Thuidium talongense* Besch.; Present status: **Synonym of** *Pelekium fuscatum* (Besch.) Touw

Distribution in India: India Orientalis

631. *Thuidium tamariscellum* (Müll. Hal.) Bosch & Sande Lac. ; **Present status: Valid name**

Distribution in India: Western Himalayas, Eastern Himalayas and South India

632. *Thuidium venustulum* Besch.; Present status: **Synonym of** *Palamocladium wilkesianum* (Sull.) Müll. Hal.

Distribution in India: Eastern Himalayas

633. *Thuidium vestitissimum* Besch.; Present status: **Synonym of** *Cyrto-hypnum vestitissimum* (Besch.) W.R. Buck & H.A. Crum

Distribution in India: Western Himalayas

Family: Hypnaceae Schimp.

115. ***Bryosedgwickia*** Cardot & Dixon

634. *Bryosedgwickia aurea* (Schwägr.) M. Fleisch. [Asthana & Sahu, 2013]; **Present status: Valid name**

Distribution in India: Western Himalayas and Eastern Himalayas

635. *Bryosedgwickia densa (Hook.)* Bizot & P. de la Varde; **Present status: Valid name**

Distribution in India: South India

636. *Bryosedgwickia kirtikarii* Cardot *et* Dixon; **Present status: Valid name**

Distribution in India: South India

637. *Bryosedgwickia polyantha* (Hedw.) Bruch & Schimp.; Present status: **Doubtful name**

Distribution in India: Western Himalayas

116. *Calliergonella* Loeske

638. *Calliergonella cuspidata* (Hedw.) Loeske; **Present status: Valid name**

Distribution in India: Eastern Himalayas

117.*Ctenidium* (Schimp.) Mitt.

639. *Ctenidium lychnites* (Mitt.) Broth.; **Present status: Valid name**

Distribution in India: Eastern Himalayas, South India, Nicobar Island

118. *Ectropothecium* Mitt.

640. *Ectropothecium buitenzorgii* (Bél.) Mitt.; **Present status: Valid name**

Distribution in India: Eastern Himalayas and Nicobar Island

641. *Ectropothecium compressifolium* (Mitt.) A. Jaeger; **Present status: Valid name**

Distribution in India: Eastern Himalayas

642. *Ectropothecium cygnicollum* (Mitt.) A. Jaeger; **Present status: Valid name**

Distribution in India: Eastern Himalayas and Nicobar Island

643. *Ectropothecium cyperoides* (Hook. & Harv.) A. Jaeger ; **Present status: Valid name**

Distribution in India: Western Himalayas, Eastern Himalayas and South India

644. *Ectropothecium cyperoides* var. *papillosum* Cardot & Dixon ; Present status: **Doubtful name**

Distribution in India: Western Himalayas, Eastern Himalayas and South India

645. *Ectropothecium dealbatum* (Reinw. & Hornsch.) A. Jaeger; **Present status: Valid name**

Distribution in India: Eastern Himalayas

646. *Ectropothecium densum* Dixon & P. de la Varde; **Present status: Valid name**

Distribution in India: South India

647. *Ectropothecium drepanocladioides* Broth. & P. de la Varde; **Present status: Valid name**

Distribution in India: South India

648. *Ectropothecium kerstanii* Dixon & Herzog; **Present status: Valid name**

Distribution in India: Eastern Himalayas (Endemic to India)

649. *Ectropothecium laevigatum* Thwaites & Mitt.; Present status: **Doubtful name**

Distribution in India: South India

650. *Ectropothecium manii* Broth.; Present status: **Doubtful name**

Distribution in India: South India and Andaman Island (Endemic to India)

651. *Ectropothecium monumentorum* (Duby) A. Jaeger; **Present status: Valid name**

Distribution in India: Eastern Himalayas

652. *Ectropothecium ramuligerum* Dixon; Present status: **Doubtful name**

Distribution in India: Eastern Himalayas (Endemic to India)

653. *Ectropothecium rostellatum* (Mitt.) A. Jaeger; **Present status: Valid name**

Distribution in India: Eastern Himalayas

654. *Ectropothecium sikkimense* (Renauld & Cardot) Renauld & Cardot ; **Present status: Valid name**

Distribution in India: Western Himalayas and Eastern Himalayas

655. *Ectropothecium stereodontoides* (Dixon) Wijk & Margad.; **Present status: Valid name**

Distribution in India: South India

119. *Gollania* Broth.

656. *Gollania clarescens* (Mitt.) Broth.; **Present status: Valid name**

Distribution in India: Western Himalayas

657. *Gollania cylindricarpa* (Mitt.) Broth.; **Present status: Valid name**

Distribution in India: India Orientalis

658. *Gollania ruginosa* (Mitt.) Broth.; **Present status: Valid name**

Distribution in India: Western Himalayas

120. *Hageniella* Broth.

659. *Hageniella assamica* Dixon [Alam, 2013]; Present status: **Doubtful name**

Distribution in India: Eastern and Western Himalayas

660. *Hageniella isopterygioides* Dixon; Present status: **Doubtful name**

Distribution in India: Eastern Himalayas (Endemic to India)

661. *Hageniella sikkimensis* Broth.; Present status: **Doubtful name**

Distribution in India: Eastern Himalayas (Endemic to India)

121. *Homomallium* (Schimp.) Loeske

662. *Homomallium incurvatum* (Schrad. & Brid.) Loeske ; **Present status: Valid name**

Distribution in India: Western Himalayas

663. *Homomallium simalaense* (Mitt.) Broth.; **Present status: Valid name**

Distribution in India: Western Himalayas

122. *Hypnum* Hedw.

664. *Hypnum aduncoides* (Brid.) Müll. Hal.; **Present status: Valid name**

Distribution in India: Eastern Himalayas

665. *Hypnum cupressiforme* Hedw.; **Present status: Valid name**

Distribution in India: Western Himalayas, Eastern Himalayas and South India

666. *Hypnum flaccens* Besch.; **Present status:** Synonym of *Hypnum macrogynum* Besch.

Distribution in India: Eastern Himalayas

667. *Hypnum macrogynum* Besch.; **Present status: Valid name**

Distribution in India: Eastern Himalayas

668. *Hypnum sikkimense* Ando; **Present status: Valid name**

Distribution in India: Eastern Himalayas

669.*Hypnum subimponens* Lesq.; **Present status: Valid name**

Distribution in India: Eastern Himalayas

670. *Hypnum subimponens* subsp. *ulophyllum* (Müll. Hal.) Ando ; **Present status: Valid name**

Distribution in India: Eastern Himalayas

671. *Hypnum submolluscum* Besch.; **Present status: Doubtful name**

Distribution in India: Eastern Himalayas

672. *Hypnum vaucheri* fo. *vaucheri* Lesq.; **Present status: Valid name**

Distribution in India: Western Himalayas

123. *Isopterygiopsis* Iwats.

673. *I. muelleriana* (Schimp.) Z. Iwats.; **Present status: Valid name**

Distribution in India: Western Himalayas and Eastern Himalayas

124. *Isopterygium* Mitt.

674. *Isopterygium albescens* var. *smallii* (Sull. & Lesq.) Z. Iwats.; **Present status: Valid name**

Distribution in India: Western Himalayas, Eastern Himalayas and South India

675. *Isopterygium andamanicum* Gangulee; **Present status: Valid name**

Distribution in India: Andaman Islands (Endemic to India)

676. *Isopterygium assamicum* (Mitt.) A. Jaeger; **Present status: Valid name**

Distribution in India: Eastern Himalayas

677. *Isopterygium bancanum* (Sande Lac.) A. Jaeger; **Present status: Valid name**

Distribution in India: Indian Orientalis

678. *Isopterygium distichaceum* (Mitt.) A. Jaeger.; Present status: **Synonym of** *Pseudotaxiphyllum distichaceum* (Mitt.) Z. Iwats.

Distribution in India: Western Himalayas and Eastern Himalayas

679. *Isopterygium elegans* (Brid.) Lindb. [Saxena & al., 2007]; Present status: **Synonym of** *Pseudotaxiphyllum elegans* (Brid.) Z. Iwats.

Distribution in India: Western Himalayas and Eastern Himalayas

680. *Isopterygium lignicola* (Mitt.) A. Jaeger; **Present status: Valid name**

Distribution in India: Eastern Himalayas and South India

681. *Isopterygium longitheca* (Mitt.) A. Jaeger; **Present status: Valid name**

Distribution in India: Eastern Himalayas (Endemic to India)

682. *Isopterygium micans* (Sw.) Kindb.; **Present status: Valid name**

Distribution in India: Eastern Himalayas

683. *Isopterygium microplumosum* (Müll. Hal.) Broth.; **Present status: Valid name**

Distribution in India: Andaman Islands

684. *Isopterygium minutirameum* (Müll. Hal.) Broth.; **Present status: Valid name**

Distribution in India: Western Himalayas, Eastern Himalayas and South India

685. *Isopterygium muellerianum* (Schimp.) A. Jaeger.; Present status: Synonym of *Isopterygiopsis muelleriana* (Schimp.) Z. Iwats.

Distribution in India: Western Himalayas

686. *Isopterygium pallidulum* (Mitt.) A. Jaeger.; **Present status: Valid name**

Distribution in India: Western Himalayas and Eastern Himalayas

687. *Isopterygium pohliaecarpum* (Sull. & Lesq.) A. Jaeger; **Present status: Valid name**

Distribution in India: Eastern Himalayas and South India

688. *Isopterygium pulchellum* (Hedw.) A. Jaeger.; Present status: Synonym of *Isopterygiopsis pulchella* (Hedw.) Z. Iwats.

Distribution in India: Western Himalayas

689. *Isopterygium seligeri* (Brid.) Dixon; Present Status: **Synonym of** *Herzogiella seligeri* (Brid.) Z. Iwats.

Distribution in India: Eastern Himalayas

690. *Isopterygium serrulatum* M. Fleisch.; **Present status: Doubtful name**

Distribution in India: Eastern Himalayas

691. *Isopterygium subalbescens* Broth.; Present Status: **Synonym of** *Isopterygium albescens* var. *smallii* (Sull. & Lesq.) Z. Iwats.

Distribution in India: Eastern Himalayas and South India

692. *Isopterygium textorii* (Sande Lac.) Mitt.; **Present status: Valid name**

Distribution in India: Eastern Himalayas and South India

693. *Isopterygium undulatum* Dixon & P. de la Varde; **Present status: Doubtful name**

Distribution in India: South India

124. *Leiodontium* Broth.

694. *Leiodontium complanatum* Dixon; Present status**: Synonym of** *Micralsopsis complanata* (Dixon) W.R. Buck

Distrbution in India: Eastern Himalayas (Endemic to India)

695. *Leiodontium gracile* Broth.; Present status: **Doubtful name**

Distribution in India: Western Himalayas

125. *Macrothamniella* M. Fleisch.

696. *Macrothamniella pilosula* (Mitt.) M. Fleisch.; **Present status: Valid name**

Distrbution in India: Eastern Himalayas

126. *Mittenothamnium* Henn.

697. *Mittenothamnium mollissimum* Cardot; Present status: **Doubtful name**

Distrbution in India: Eastern Himalayas

127. *Nanothecium* Dixon & P. de la Varde

698. *Nanothecium foreaui* Dixon & P. de la Varde; Present status: **Doubtful name**

Distrbution in India: South India

128. *Orthothecium* Schimp.

699. *Orthothecium intricatum* (Hartm.) Schimp.; Present status: **Doubtful name**

Distrbution in India: Western Himalayas and Punjab

129. *Platydictya* Berk.

700. *Platydictya jungermannioides* (Brid.) H.A. Crum ; **Present status: Valid name**

Distrbution in India: Western Himalayas

701. *Platydictya madurensis* (Cardot & P. de la Varde) R.S. Chopra; **Present status: Valid name**

Distribution in India: South India

702. *Platydictya subtilis* (Hedw.) H. A. Crum; **Present status: Valid name**

Distrbution in India: Western Himalayas

130. *Platygyrium* Bruch & Schimp.

703. *Platygyrium repens* (Brid.) Schimp.; **Present status: Valid name**

Distrbution in India: Western Himalayas

704. *Platygyrium russulum* (Mitt.) A. Jaeger ; Present status**: Synonym of** *Gammiella russula* (Mitt.) M. Fleisch.

Distrbution in India: Western Himalayas and Eastern Himalayas

705. *Platygyrium subrussulum* Renauld & Cardot ; **Present status: Doubtful name**

Distribution in India: Eastern Himalayas (Endemic to India)

131. *Pleuridium* Brid.

706. *Pleuridium denticulatum* (Müll. Hal.) Mitt.; **Present status: Valid name**

Distribution in India: South India

707. *Pleuridium mussuriense* Broth. & G. Roth ; *Present status*: **Synonym of** *Brothera leana* (Sull.) Müll. Hal.

Distrbution in India: Western Himalayas

708. *Pleuridium tenue* Mitt.; Present status**: Synonym of** *Astomum tenue* (Mitt.) Müll. Hal.

Distrbution in India: Western Himalayas and Eastern Himalayas

132. *Pseudostereodon* (Broth.) M. Fleisch.

709. *Pseudostereodon procerrimus* (Molendo) M. Fleisch.; Present status: **Synonym** is a Synonym of *Ctenidium procerrimum* (Molendo) Lindb.

Distrbution in India: Western Himalayas

133. *Pseudosymblepharis* Broth.

710. *Pseudosymblepharis angustata* (Mitt.) Hilp.; **Present status: Valid name**

Distrbution in India: Western Himalayas and South India

711. *Pseudosymblepharis pallidens* Dixon; Present status: **Synonym of** *Pseudosymblepharis bombayensis* (Müll. Hal.) P. Sollman

Distribution in India: Eastern Himlayas (Endemic to India).

712. *Pseudosymblepharis subduriuscula* (Müll. Hal.) P.C. Chen ; Present status: **Synonym of** *Pseudosymblepharis angustata* (Mitt.) Hilp.

Distrbution in India: Western Himalayas

Family: Lembophyllaceae Broth.

134. *Neobarbella* Nog.

713. *Neobarbella comes* (Griff.) Nog.; **Present status: Valid name**

Distribution in India: Eastern Himalyas

Family: Myriniaceae Schimp.

135. *Schwetschkeopsis* Broth.

714. *Schwetschkeopsis formosana* Nog.: **Present status: Doubtful**

Distribution in India: Eastern Himalyas

715. *Schwetschkeopsis neckeroides* (Griff.) Vohra: Present status: **Synonym of** *Levierella neckeroides* (Griff.) O'Shea & Matcham

Distribution in India: Eastern Himalyas

Family Pterigyandraceae Schimp.

136. *Pterigynandrum* Hedw.

716. *Pterigynandrum decolor* (Mitt.) Broth.; Present status: **Synonym of** *Leptopterigynandrum decolor* (Mitt.) M. Fleisch.

Distribution in India: Eastern Himalyas

717. *Pterigynandrum filiforme* Hedw.; **Present status: Valid name**

Distrbution in India: Western Himalayas

137. *Ptilium* De Not.

718. *Ptilium crista-castrensis* (Hedw.) De Not.; **Present status: Valid name**

Distribution in India: Eastern and Western Himalayas

138. *Pylaisia* Bruch & Schimp.

719. *Pylaisia brevirostris* (Griff.) A. Jaeger; **Present status: Valid name**

Distribution in India: Eastern Himalayas (Endemic to India)

720. *Pylaisia extenta* A. Jaeger; Present status: **Synonym of** *Pylaisiella extenta* (Mitt.) Ando

Distribution in India: Eastern Himalayas

721. *Pylaisia falcata* Thér.; Present status: **Synonym of** *Pylaisiella falcata (Schimp.) Ando*

Distribution in India: Eastern Himalayas

722. *Pylaisia polyantha* (Hedw.) Schimp.; Present status: **Synonym of** *Pylaisiella polyantha* (Hedw.) Grout

Distribution in India: Western Himalayas

139. *Taxiphyllum* M. Fleisch.

723. *Taxiphyllum giraldii* (Müll. Hal.) M. Fleisch.; **Present status: Valid name**

Distribution in India: Western Himalayas

724. *Taxiphyllum maniae* (Renauld & Paris) M. Fleisch.; Present status: **Synonym of** *Taxiphyllum taxirameum* (Mitt.) M. Fleisch.

Distribution in India: Western Himalayas and Nicobar Islands

725. *Taxiphyllum taxirameum* (Mitt.) M. Fleisch.; **Present status: Valid name**

Distribution in India: Western Himalayas, Eastern Himalayas, South India, Central India and Nicobar Islands

140. *Vesicularia* (Müll. Hal.) Müll. Hal.

726. *Vesicularia firma* Dixon & P. de la Varde; Present status: **Doubtful name**

Distribution in India: South India and Central India

727. *Vesicularia kurzii* (A. Jaeger) Broth.; **Present status: Valid name**

Distribution in India: South India and Central India

728. *Vesicularia montagnei* (Schimp.) Broth.; **Present status: Valid name**

Distribution in India: Western Himalayas, Eastern Himalayas and Gangetic plains

729. *Vesicularia nitidula* Cardot & P. de la Varde; Present status: **Doubtful name**

Distribution in India: South India

730. *Vesicularia perreticulata* Broth. ex Dixon; Present status: **Doubtful name**

Distribution in India: South India

731. *Vesicularia reticulata* (Dozy & Molk.) Broth.; **Present status: Valid name**

Distribution in India: Western Himalayas, Eastern Himalayas and South India

732. *Vesicularia subpilicuspis* Cardot & P. de la Varde; Present status: **Doubtful name**

Distribution in India: South India

733. *Vesicularia subscaturiginosa* M. Fleisch.; Present status: **Doubtful name**

Distribution in India: Western Himalayas

734. *Vesicularia succosa* (Mitt.) Broth.; **Present status: Valid name**

Distribution in India: Western Himalayas, Eastern Himalayas and Central India

Family: Pterobryaceae Kindb.

141. *Calyptothecium* Mitt.

735. *Calyptothecium dixonii* Gangulee; Present status: **Synonym of** *Calyptothecium wightii* (Mitt.) M. Fleisch.

Distribution in India: Eastern Himalayas (Endemic to India)

736. *Calyptothecium himantocladioides* E.B. Bartram; **Present status: Valid name**

Distribution in India: Eastern Himalayas

737. *Calyptothecium hookeri* (Mitt.) Broth.; **Present status: Valid name**

Distribution in India: Western Himalayas and Eastern Himalayas

738. *Calyptothecium nitidum* (Renauld & Cardot) M. Fleisch.; **Present status: Valid name**

Distribution in India: Eastern Himalayas (Endemic to India)

739. *Calyptothecium patulum* (Broth.) M. Fleisch.; **Present status: Valid name**

Distribution in India: South India

740. *Calyptothecium pinnatum* Nog.; **Present status: Valid name**

Distribution in India: Eastern Himalayas

741. *Calyptothecium symphysodontoides* Dixon & P. de la Varde; **Present status: Doubtful**

Distribution in India: South India

742. *Calyptothecium urvilleanum* (Müll. Hal.) Broth.; **Present status: Valid name**

Distribution in India: Eastern Himalayas and South India

743. *Calyptothecium wightii* (Mitt.) M. Fleisch.; **Present status: Valid name**

Distribution in India: Eastern Himalayas

142. *Garovaglia* Endl.

744. *Garovaglia plicata* (Brid.) Bosch & Sande Lac.; **Present status: Valid name**

Distribution in India: Eastern Himalayas

143. *Horikawaea* Nog.

745. *Horikawaea nitida* Nog.; **Present status: Doubtful name**

Distribution in India: Eastern Himalayas

144. *Jaegerina* C. Muell.

746. *Jaegerina stolonifera* (Müll. Hal.) Müll. Hal.; **Present status: Valid name**

Distribution in India: South India

Family: Stereophyllaceae (M. Fleisch.) W. R. Buck & Ireland.

145. *Juratzekaea* Lorentz

747. *Juratzkaea indica* Broth. & P. de la Varde; **Present status: Doubtful name**

Distribution in India: South India

146. *Myurium* Schimp.

748. *Myurium perplexum* (Renauld & Cardot) Broth.; **Present status: Valid name**

Distribution in India: Eastern Himalayas

749. *Myurium rufescens* (Reinw. & Hornsch.) M. Fleisch.; **Present status: Valid name**

Distribution in India: Eastern Himalayas and South India

750. *Myurium warburgii* (Müll. Hal.) M. Fleisch.; **Present status: Valid name**

Distribution in India: South India

147. *Osterwaldiella* Fleisch. ex Broth.

751. *Osterwaldiella monostricta* M. Fleisch. ex Broth.; **Present status: Doubtful name**

Distribution in India: Eastern Himalayas

148. *Symphysodontella* Fleisch.

752. *Symphysodontella involuta* (Thwaites & Mitt.) M. Fleisch.; **Present status: Valid name**

Distribution in India: South India

753. *Symphysodontella pilifolia* Dixon; **Present status:** Synonym of *Pterobryopsis pilifolia* (Dixon) Magill

Distribution in India: Eastern Himalayas

754. *Symphysodontella subulata* Broth.; **Present status: Valid name**

Distribution in India: Eastern Himalayas

755. *Symphysodontella tortifolia* Dixon; **Present status: Doubtful**

Distribution in India: Eastern Himalayas

Family: Entodontaceae Kindb.

149. ***Entodon*** Müll. Hal.

756. *Entodon chloropus* Renauld & Cardot; **Present status: Valid name**

Distribution in India: Eastern Himalayas

757. *Entodon concinnus* (De Not.) Paris; **Present status: Valid name**

Distribution in India: Indian Orienatalis

758. *Entodon concinnus* subsp. *caliginosus* (Mitt.) Mizush.; **Present status: Valid name**

Distribution in India: Indian Orientalis

759. *Entodon curvatus* (Griff.) A. Jaeger; **Present status: Synonym of** *Cylindrothecium curvatum* (Griff.) Paris

Distribution in India: Eastern Himalayas and South India (Endemic to India)

760. *Entodon flavescens* (Hook.) A. Jaeger ; **Present status: Valid name**

Distribution in India: Western Himalayas

761. *Entodon laetus* (Griff.) A. Jaeger; **Present status: Synonym of** *Cylindrothecium laetum* (Griff.) Paris

Distribution in India: Eastern Himalayas and South India

762. *Entodon longifolius* (Müll. Hal.) A. Jaeger; **Present status: Synonym of** *Cylindrothecium longifolium* (Müll. Hal.) Paris

Distribution in India: South India

763. *Entodon luridus* (Griff.) A. Jaeger; **Present status: Valid name**

Distribution in India: Eastern Himalayas and South India

764. *Entodon luteonitens* Renauld & Cardot; **Present status: Doubtful name**

Distribution in India: Eastern Himalayas (Endemic to India)

765. *Entodon macropodus* (Hedw.) Müll. Hal.; **Present status: Doubtful name**

Distribution in India: Eastern Himalayas (Endemic to India)

766. *Entodon myurus* (Hook.) Hampe ; *Present status*: **Synonym of** *Cylindrothecium myurum* (Hook.) Paris

Distribution in India: Western Himalayas and Eastern Himalayas

767. *Entodon obtusatus* Broth.; **Present status: Valid name**

Distribution in India: South India

768. *Entodon ovicarpus* Dixon; **Present status: Doubtful name**

Distribution in India: Eastern Himalayas

769. *Entodon plicatus* Müll. Hal.; **Present status: Valid name**

Distribution in India: South India

770. *Entodon prorepens* (Mitt.) A. Jaeger.; **Present status: Valid name**

Distribution in India: Western Himalayas, Eastern Himalayas and South India

771. *Entodon pulchellus* (Griff.) A. Jaeger; *Present status*: **Synonym of** *Cylindrothecium pulchellum* (Griff.) Paris

Distribution in India: Eastern Himalayas (Endemic to India)

772. *Entodon rubicundus* (Mitt.) A. Jaeger.; **Present status: Valid name**

Distribution in India: Western Himalayas, Eastern Himalayas and South India

773. *Entodon plicatus* Müll. Hal.; **Present status: Valid name**

Distribution in India: Western Himalayas, Eastern Himalayas, Central India, Gangetic plains and South India

774. *Entodon scariosus* Renauld & Cardot; *Present status*: **Synonym of** *Cylindrothecium scariosum* (Renauld & Cardot) Paris

Distribution in India: Eastern Himalayas

775. *Entodon subplicatus* Renauld & Cardot ; Present status: **Doubtful name**

Distribution in India: Western Himalayas and Eastern Himalayas

150. *Endotrichella* C. Muell.

776.*Endotrichella eberhardtii* Broth. & Paris; **Present status: Synonym of** *Garovaglia eberhardtii* (Broth. & Paris) During

777. *Endotrichella elegans* (Dozy & Molk.) M. Fleisch.; **Present status: Valid name**

Distribution in India: Eastern Himalayas

778. *Endotrichella poilaneana* Thér. & P. de la Varde; **Present status: Synonym of** *Garovaglia elegans* (Dozy & Molk.) Hampe ex Bosch & Sande Lac.

Distribution in India: South India

151. *Penzigiella* M. Fleisch.

779. *Penzigiella cordata* (Hook. & Harv.) M. Fleisch.; **Present status: Valid name**

Distribution in India: Western Himalayas, Eastern Himalayas and South India

152. *Pterobryopsis* M. Fleisch.

780. *Pterobryopsis acuminata* (Hook.) M. Fleisch.; **Present status: Valid name**

Distribution in India: Eastern Himalayas

781. *Pterobryopsis auriculata* Dixon; **Present status: Synonym of** *Calyptothecium auriculatum* (Dixon) Nog.

Distribution in India: Eastern Himalayas and South India

782. *Pterobryopsis conchophylla* (Renauld & Cardot) Broth.; **Present status: Valid name**

Distribution in India: Eastern Himalayas

783. *Pterobryopsis divergens* (Mitt.) Nog.; **Present status: Valid name**

Distribution in India: Eastern Himalayas

784. *Pterobryopsis flexipes* (Mitt.) M. Fleisch.; **Present status: Valid name**

Distribution in India: Eastern Himalayas and South India

785. *Pterobryopsis nematosa* (Müll. Hal.) Broth. ex Paris; **Present status: Valid name**

Distribution in India: Eastern Himalayas

786. *Pterobryopsis orientalis* (Müll. Hal.) M. Fleisch.; **Present status: Valid name**

Distribution in India: Western Himalayas, Eastern Himalayas and South India

787. *Pterobryopsis schmidii* (Müll. Hal.) M. Fleisch.; **Present status: Valid name**

Distribution in India: South India

Family: Brachyteciaceae G. Roth.

153. ***Aerobryum*** Dozy & Molk.

788. *Aerobryum speciosum* Dozy & Molk.; **Present status: Synonym of** *Meteorium speciosum* (Dozy & Molk.) Mitt. (family Meteoriaceae).

Distribution in India: Eastern Himalayas

789. *Aerobryum willisii* M. Fleisch.; **Present status: Synonym of** of *Pterobryopsis tumida* (Dicks. ex Hook.) Dixon. (family **Pterobryaceae** Kindb.)

Distribution in India: South India

154. ***Brachythecium*** Schimp.

790. *Brachythecium brachycladum* (Broth.) Paris ; Present status: **Doubtful name**

Distribution in India: Western Himalayas (Endemic to India)

791. *Brachythecium buchananii* (Hook.) A. Jaeger.; **Present status: Valid name**

Distribution in India: Western Himalayas, Eastern Himalayas and South India

792. *Brachythecium buchananii* var. *cuspidiferum* (Mitt.) Gangulee & Vohra ; **Present status: Valid name**

Distribution in India: Western Himalayas, Eastern Himalayas and South India

793. *Brachythecium cameratum* (Mitt.) A. Jaeger ; *Present status*: **Synonym of** *Cirriphyllum cameratum* (Mitt.) Broth.

Distribution in India: Western Himalayas and Eastern Himalayas

794. *Brachythecium campestre* (Müll. Hal.) Schimp.; **Present status: Valid name**

Distribution in India: Western Himalayas

795. *Brachythecium chakratense* Vohra ; Present status: **Doubtful name**

Distribution in India: Western Himalayas

796. *Brachythecium cirrosum* (Schwägr.) Schimp.; **Present status: Valid name**

Distribution in India: Western Himalayas

797. *Brachythecium curvatulum* (Broth.) Paris ; **Present status: Valid name**

Distribution in India: Western Himalayas (Endemic to India)

798. *Brachythecium falcatulum* (Broth.) Paris v **Present status: Valid name**

Distribution in India: Western Himalayas

799. *Brachythecium formosanum* Takaki; **Present status: Doubtful name**

Distribution in India: Eastern Himalayas

800. *Brachythecium garhwalense* Vohra ; Present status: **Doubtful name**

Distribution in India: Western Himalayas (Endemic to India)

801. *Brachythecium glaciale* Schimp.; **Present status: Valid name**

Distribution in India: Western Himalayas

802. *Brachythecium indicopopuleum* Dixon ; Present status: **Doubtful name**

Distribution in India: Western Himalayas (Endemic to India)

803. *Brachythecium kamounense* (Harv.) A. Jaeger.; **Present status: Valid name**

Distribution in India: Western Himalayas and Eastern Himalayas

804. *Brachythecium kashmirense* (Broth.) Paris ; Present status**: Synonym of** *Brachytheciastrum kashmirense* (Broth.) Ignatov & Huttunen

Distribution in India: Western Himalayas (Endemic to India)

805. *Brachythecium laevivelutinum* Dixon ; Present status: **Doubtful name**

Distribution in India: Western Himalayas (Endemic to India)

806. *Brachythecium laxifolium* Dixon ; Present status: **Doubtful name**

Distribution in India: Western Himalayas (Endemic to India)

807. *Brachythecium longicuspidatum* (Mitt.) A. Jaeger.; **Present status: Valid name**

Distribution in India: Western Himalayas (Endemic to India)

808. *Brachythecium microsericeum* Dixon; **Present status: Valid name**

Distribution in India: Western Himalayas and Eastern Himalayas(Endemic to India)

809. *Brachythecium myurelliforme* Dixon**;** Present status**: Synonym of** *Brachythecium collinum* (Schleich. & Müll. Hal.) Schimp.

Distribution in India: Western Himalayas (Endemic to India)

810. *Brachythecium nitidissimum* Dixon & P. de la Varde; Present status: **Doubtful name**

Distribution in India: South India

811. *Brachythecium obsoletinerve* Dixon ; Present status: **Doubtful name**

Distribution in India: Western Himalayas (Endemic to India)

812. *Brachythecium oedistegum* (Müll. Hal.) A. Jaeger; **Present status: Valid name**

Distribution in India: South India

813. *Brachythecium pachytheca* (Dixon) Vohra; **Present status: Valid name**

Distribution in India: India Orientalis

814. *Brachythecium plumosum* (Hedw.) Bruch & Schimp. [Asthana & Sahu, 2013]; Present status: **Doubtful name**

Distribution in India: Western Himalayas, Eastern Himalayas and South India

815. *Brachythecium populeum* (Hedw.) Bruch & Schimp.; **Present status: Valid name**

Distribution in India: Western Himalayas

816. *Brachythecium procumbens* (Mitt.) A. Jaeger.; **Present status: Valid name**

Distribution in India: Western Himalayas, Eastern Himalayas and South India

817. *Brachythecium reflexum* (Starke) Schimp.; **Present status: Valid name**

Distribution in India: India Orientalis

818. *Brachythecium rivulare* Bruch & Schimp. ; **Present status: Valid name**

Distribution in India: Western Himalayas and Eastern Himalayas

819. *Brachythecium rivulare* var. *alare* (Dixon) Vohra ; **Present status: Valid name**

Distribution in India: Western Himalayas (Endemic to India)

820. *Brachythecium rutabulum* (Hedw.) Schimp.; **Present status: Valid name**

Distribution in India: Western Himalayas

821. *Brachythecium salebrosum* (Hoffm. et F. Weber ex D. Mohr) Schimp.; **Present status: Valid name**

Distribution in India: Western Himalayas

822. *Brachythecium spurio-populeum* (Broth.) Paris ; **Present status: Valid name**

Distribution in India: Western Himalayas (Endemic to India)

823. *Brachythecium starkii* var. *explanatum* (Brid.) Mönk.; *Present status*: **Synonym of** *Brachythecium oedipodium* (Mitt.) A. Jaeger

Distribution in India: Western Himalayas

824. *Brachythecium stricticalyx* Dixon & Badhw.; Present status: **Doubtful name**

Distribution in India: Western Himalayas

825. *Brachythecium velutinum* (Hedw.) Schimp.; **Present status: Valid name**

Distribution in India: Western Himalayas

826. *Brachythecium waziriense* Dixon ; Present status: **Doubtful name**

Distribution in India: Western Himalayas

827. *Brachythecium wichurae* (Broth.) Paris; Present status: **Synonym of** *Brachythecium grovaglioides* Müll. Hal.

Distribution in India: Eastern Himalayas

155. *Bryhnia* Kaur.

828. *Bryhnia decurvans* (Mitt.) Dixon ; **Present status: Valid name**

Distribution in India: Western and Eastern Himalyas

829. *Bryhnia novae-angliae* (Sull. & Lesq.) Grout.; **Present status: Valid name**

Distribution in India: Eastern Himalyas

156. *Camptothecium* Bruch & Schimp.

830. *Camptothecium lutescens* (Hedw.) Schimp.; **Present status: Valid name**

Distribution in India: Western Himalayas

157. *Cirriphyllum* Grout

831. *Cirriphyllum alare* Dixon ; Present status: **Doubtful name**

Distribution in India: Western Himalayas

832. *Cirriphyllum cameratum* (Mitt.) Broth.; **Present status: Valid name**

Distribution in India: Western Himalayas and Eastern Himalayas

833. *Cirriphyllum cirrosum* (Schwägr.) Grout ; **Present status: Valid name**

Distribution in India: Western Himalayas

158. *Eurhynchium* Bruch & Schimp.

834. *Eurhynchium dumosum* (Mitt.) A. Jaeger; **Present status: Valid name**

Distribution in India: Eastern Himalayas

835. *Eurhynchium hians* (Hedw.) Sande Lac.; **Present status: Valid name**

Distribution in India: Western Himalayas and Eastern Himalayas

836. *Eurhynchium muelleri* (A. Jaeger.) E. B. Bartram ; **Present status: Valid name**

Distribution in India: Western Himalayas and Eastern Himalayas

837. *Eurhynchium muelleri* var. *minus* Dixon ; Present status: **Doubtful name**

Distribution in India: Western Himalayas and Eastern Himalayas

838. *Eurhynchium ovatum* Cardot.; Present status: **Doubtful name**

Distribution in India: South India

839. *Eurhynchium praelongum* (Hedw.) Schimp.; **Present status: Valid name**

Distribution in India: Western Himalayas and Eastern Himalayas

840. *Eurhynchium pulchellum* (Hedw.) Jenn.; **Present status: Valid name**

Distribution in India: Western Himalayas and Eastern Himalayas

841. *Eurhynchium riparioides* (Hedw.) P. W. Richards ; **Present status: Valid name**

Distribution in India: Western Himalayas, Eastern Himalayas and South India

842. *Eurhynchium striatum* (Schreb. & Hedw.) Schimp.; **Present status: Valid name**

Distribution in India: Western Himalayas

843. *Eurhynchium swartzii* (Turner) Curn.; *Present status*: **Synonym of** *Eurhynchium hians* (Hedw.) Sande Lac.

Distribution in India: Eastern Himalayas

159. *Homalothecium* Bruch & Schimp.

844. *Homalothecium euchloron* (Bruch & Müll. Hal.) R.S. Chopra ; **Present status: Valid name**

Distribution in India: Western Himalayas

845. *Homalothecium incompletum* (Griff.) A. Jaeger; **Present status: Valid name**

Distribution in India: Eastern Himalayas (Endemic to India)

846. *Homalothecium integerrimum* Dixon ; Present status: **Doubtful name**

Distribution in India: Western Himalayas (Endemic to India)

847. *Homalothecium neckeroides* (Griff.) Paris; **Present status: Valid name**

Distribution in India: Eastern Himalayas

848. *Homalothecium nilgheriense* (Mont.) H. Rob.; **Present status: Valid name**

Distribution in India: Western Himalayas, Eastern Himalayas and South India

849. *Homalothecium sericeum* (Hedw.) Schimp.; **Present status: Valid name**

Distribution in India: Western Himalayas

160. *Isothecium* Brid.

850. *Isothecium comatum* (Müll. Hal.) Hook. & Wilson; **Present status:** a Synonym of *Hypnodendron comatum* (Müll. Hal.) Touw

Distribution in India: Eastern Himalayas

161. *Palamocladium* (Müll. Hal.) H. Rob.

851. *Palamocladium euchloron* (Bruch & Müll. Hal.) Wijk & Margad.; **Present status: Valid name**

Distribution in India: Western Himalayas

852. *Palamocladium nilgheriense* (Mont.) Müll. Hal.; *Present status*: **Synonym of** *Palamocladium leskeoides* (Hook.) E. Britton

Distribution in India: Western Himalayas, Eastern Himalayas and South India

162. *Rhynchostegiella* (Bruch & Schimp.) Limpr.

853. *Rhynchostegiella assamica* Cardot & Dixon **Present status: Valid name**

Distribution in India: Eastern Himalayas (Endemic to India)

854. *Rhynchostegiella compacta* (Hook.) Loeske; **Present status: Valid name**

Distribution in India: Western Himalayas

855. *Rhynchostegiella divaricatifolia* (Renauld & Cardot) Broth. [Saxena & al., 2010]; **Present status: Valid name**

Distribution in India: Eastern Himalayas (Endemic to India)

856. *Rhynchostegiella fabroniadelphus* (Müll. Hal.) Broth.; **Present status: Valid name**

Distribution in India: South India

857. *Rhynchostegiella humillima* (Mitt.) Broth.; **Present status: Valid name**

Distribution in India: Western Himalayas, Eastern Himalayas and South India

858. *Rhynchostegiella leiopoda* Dixon & Cardot; Present status: **Doubtful name**

Distribution in India: Eastern Himalayas (Endemic to India)

859. *Rhynchostegiella menadensis* (Sande Lac.) E. B. Bartram; **Present status: Valid name**

Distribution in India: Eastern Himalayas

860. *Rhynchostegiella percomplanata* Dixon; *Present status*: **Synonym of** *Rhynchostegiella menadensis* (Sande Lac.) E.B. Bartram

Distribution in India: Eastern Himalayas

861. *Rhynchostegiella ramicola* (Broth.) Broth.; *Present status*: **Synonym of** *Remyella ramicola* (Broth.) Ignatov & Huttunen

Distribution in India: Eastern Himalayas

862. *Rhynchostegiella sachensis* Dixon; Present status: **Doubtful name**

Distribution in India: Western Himalayas (Endemic to India)

863. *Rhynchostegiella scabriseta* (Schwägr.) Broth.; **Present status: Valid name**

Distribution in India: Western Himalayas and Eastern Himalayas

163. *Rhynchostegium* Bruch & Schimp.

864. *Rhynchostegium calderii* Vohra; Present status: **Doubtful name**

Distribution in India: Eastern Himalayas

865. *Rhynchostegium celebicum* (Sande Lac.) A. Jaeger.; **Present status: Valid name**

Distribution in India: Western Himalayas, Eastern Himalayas and South India

866. *Rhynchostegium duthiei* Müll. Hal. & Dixon ; Present status: **Doubtful name**

Distribution in India: Western Himalayas, Eastern Himalayas and South India

868. *Rhynchostegium herbaceum* (Mitt.) A. Jaeger.; **Present status: Valid name**

Distribution in India: Western Himalayas, Eastern Himalayas and South India

869. *Rhynchostegium hookeri* A. Jaeger.; **Present status: Valid name**

Distribution in India: Western Himalayas and Eastern Himalayas

870. *Rhynchostegium muelleri* A. Jaeger ; Present status: **Synonym of** *Eurhynchium muelleri* (A. Jaeger) E.B. Bartram

Distribution in India: Western Himalayas, Eastern Himalayas and South India

871. *Rhynchostegium planiusculum* (Mitt.) A. Jaeger.; **Present status: Valid name**

Distribution in India: Western Himalayas and Eastern Himalayas

872. *Rhynchostegium riparioides* (Hedw.) Cardot ; **Present status: Valid name**

Distribution in India: Western Himalayas, Eastern Himalayas and South India

873. *Rhynchostegium subrectocarpum* (Dixon) Vohra ; **Present status: Valid name**

Distribution in India: Western Himalayas (Endemic to India)

874. *Rhynchostegium vagans* A. Jaeger; Present status: **Synonym of** *Oxyrrhynchium vagans* (A. Jaeger) Ignatov & Huttunen

Distribution in India: Western Himalayas, Eastern Himalayas and South India

164. *Rozea* Besch.

875. *R. pterogonioides* (Harv.) A. Jaeger.; **Present status: Valid name**

Family: Fabroniaceae Schimp.

165. *Fabronia* Raddi

876. *Fabronia assamica* Dixon; Present status: **Doubtful name**

Distribution in India: Eastern Himalayas (Endemic to India)

877. *Fabronia ciliaris* (Brid.) Brid.; **Present status: Valid name**

Distribution in India: Western Himalayas

878. *Fabronia curvirostris* Dozy & Molk.; Present status: **Doubtful name**

Distribution in India: South India

879. *Fabronia goughi* Mitt.; Present status: **Doubtful name**

Distribution in India: Western Himalayas and South India

880. *Fabronia madurensis* Dixon & P. de la Varde ; Present status: **Doubtful name**

Distribution in India: Western Himalayas and South India

881. *Fabronia minuta* Mitt.; Present status: **Doubtful name**

Distribution in India: Western Himalayas and Central India

882. *Fabronia pusilla* Raddi; Present status: **Valid name**

Distribution in India: India Orientalis

883. *Fabronia schmidii* Müll. Hal.; Present status: **Doubtful name**

Distribution in India: South India

884. *Fabronia secunda* Mont. [Asthana & Sahu, 2007]; **Present status: Valid name**

Distribution in India: South India and Eastern Himalayas

166. *Levierella* Müll. Hal.

885. *Levierella fabroniacea* Müll. Hal.; Present status**: Synonym of** *Levierella neckeroides* (Griff.) O'Shea & Matcham

Distribution in India: Western Himalayas, Eastern Himalayas, South India and Central India

Family: Meteoriaceae Kindb.

167. *Aerobryopsis* M. Fleisch.

886 *Aerobryopsis denticulatum* Dixon; Present status**: Synonym of** *Barbella compressiramea* (Renauld & Cardot) M. Fleisch.

Distribution in India: Eastern Himalayas

887. *Aerobryopsis deflexa* Broth. & Paris; **Present status: Doubtful name**

Distribution in India: South India

888. *Aerobryopsis denticulata* Dixon: **Present status: Valid name**

Distribution in India: Eastern Himalayas

889. *Aerobryopsis longissima* (Dozy & Molk.) M. Fleisch.; Present status**: Synonym of** *Aerobryopsis wallichii* (Brid.) M. Fleisch.

Distribution in India: Western Himalayas, Eastern Himalayas and South India

890. *Aerobryopsis membranacea* (Mitt.) Broth.; **Present status: Valid name**

Distribution in India: Eastern Himalayas

891. *Aerobryopsis subleptostigmata* Broth. & Paris; **Present status: Doubtful name**

Distribution in India: South India

892. *Aerobryopsis wallichi* (Brid.) M. fleisch.; **Present status: Doubtful name**

Distribution in India: Eastern Himalayas

168. *Aerobryidium* M. Fleisch.

893. *Aerobryidium attenuatum* (Thwaites & Mitt.) M. Fleisch.; **Present Status:** Synonym of *Pseudobarbella attenuata* (Thwaites & Mitt.) Nog.

Distribution in India: South India

894. *Aerobryidium filamentosum* (Hook.) M. Fleisch.; **Present status: Valid name**

Distribution in India: Western Himalayas, Eastern Himalayas and South India

895. *Aerobryidium punctulatum* (Müll. Hal.) Dixon; **Present status: Valid name**

Distribution in India: South India

169. *Barbella* M. Fleisch. in Broth.

896. *Barbella angustifolia* Broth. ex Gangulee; **Present status: Doubtful name**

Distribution in India: Eastern Himalayas (Endemic to India)

897. *Barbella angustifolia* var. *subtenax* Gangulee; **Present status: Doubtful name**

Distribution in India: Eastern Himalayas (Endemic to India)

898. *Barbella bombycina* (Renauld & Cardot) M. Fleisch ; **Present status: Valid name**

Distribution in India: Western Himalayas and Eastern Himalayas

899. *Barbella compressiramea* (Renauld & Cardot) M. Fleisch ; **Present status: Valid name**

Distribution in India: Western Himalayas and Eastern Himalayas

900. *Barbella convolvens* (Mitt.) Broth.; **Present status: Valid name**

Distribution in India: Eastern Himalayas and South India

901. *Barbella cubensis* (Mitt.) Broth.; **Present status: Valid name**

Distribution in India: Eastern Himalayas and South India

902. *Barbella determesii* (Renauld & Cardot) M. Fleisch.; **Present status: Valid name**

Distribution in India: South India

903. *Barbella enervis* (Thwaites & Mitt.) M. Fleisch.; **Present status: Valid name**

Distribution in India: Eastern Himalayas and South India

904. *Barbella flagellifera* (Cardot) Nog.; **Present status: Valid name**

Distribution in India: South India

905. *Barbella pendula* (Sull.) M. Fleisch.; **Present status: Valid name**

Distribution in India: Eastern Himalayas and South India

906. *Barbella questei* Cardot & Dixon; **Present status: Synonym of** *Barbella enervis* (Thwaites & Mitt.) M. Fleisch.

Distribution in India: South India

907. *Barbella rufifolia* (Thwaites & Mitt.) Broth.; **Present status: Valid name**

Distribution in India: Eastern Himalayas and South India

908. *Barbella spiculata* (Mitt.) Broth.; **Present status: Valid name**

Distribution in India: Eastern Himalayas and South India

909. *Barbella stevensii* (Renauld & Cardot) M. Fleisch.; **Present status: Valid name**

Distribution in India: Eastern Himalayas

910. *Barbella tenax* (Müll. Hal.) Broth.; **Present status: Valid name**

Distribution in India: South India

911. *Barbella turgida* Nog.; **Present status: Doubtful name**

Distribution in India: Eastern Himalayas

171. *Cryptopapillaria* M. Menzel

912. *Cryptopapillaria fuscescens* (Hook.) M. Menzel; **Present status: Valid name**

Distribution in India: South India

172. *Chrysocladium* M. Fleisch.

913. *Chrysocladium flammeum* (Mitt.) M. Fleisch.; **Present status: Valid name**

Distribution in India: Eastern Himalayas

914. *Chrysocladium horridum* Dixon; Present status**: Synonym of** *Chrysocladium flammeum* (Mitt.) M. Fleisch.

Distribution in India: Eastern Himalayas

915. *Chrysocladium infuscatum* (Mitt.) M. Fleisch.; **Present status: Valid name**

Distribution in India: Eastern Himalayas

916. *Chrysocladium phaeum* (Mitt.) M. Fleisch.; **Present status: Valid name**

Distribution in India: Eastern Himalayas

917. *Chrysocladium retrorsum* (Mitt.) M. Fleisch.; **Present status: Valid name**

Distribution in India: South India

918. *Chrysocladium tumido-aureum* (Renauld & Cardot) M. Fleisch.; **Present status: Valid name**

Distribution in India: Eastern Himalayas

173. *Diaphanodon* Renuald & Cardot

919. *Diaphanodon blandus* (Harv.) Renuald & Cardot ; **Present status: Valid name**

Distribution in India: Western Himalayas, Eastern Himalayas, South India and Gangetic plains

920. *Diaphanodon procumbens* (Müll. Hal.) Renuald & Cardot ; **Present status: Valid name**

Distribution in India: Western Himalayas, Eastern Himalayas, South India and Gangetic plains

174. *Duthiella* Müll. Hal.

921. *Duthiella declinata* (Mitt.) Zanten ; **Present status: Valid name**

Distribution in India: Western Himalayas

922. *Duthiella flaccida* (Cardot) Broth.; **Present status: Valid name**

Distribution in India: Western Himalayas

923. *Duthiella formosana* Nog.; **Present status: Doubtful name**

Distribution in India: Eastern Himalayas

924. *Duthiella wallichii* (Mitt.) Müll. Hal.; **Present status: Valid name**

Distribution in India: Western Himalayas and Eastern Himalayas

175. *Floribundaria* M. Fleisch.

925. *Floribundaria armata* Broth.; Present status: **Synonym of** *Floribundaria setschwanica* Broth.

Distribution in India: Eastern Himalyas

926. *Floribundaria aurea* (Mitt.) Broth.; **Present status: Valid name**

Distribution in India: Eastern Himalayas

927. *Floribundaria chloronema* (Müll. Hal.) Broth.; **Present status: Valid name**

Distribution in India: Eastern Himalayas and South India

928. *Floribundaria chrysonema* (Müll. Hal.) Broth.; **Present status: Valid name**

Distribution in India: Eastern Himalayas

929. *Floribundaria commutata* (Mitt.) Broth.; **Present status: Valid name**

Distribution in India: Eastern Himalayas

930. *Floribundaria floribunda* (Dozy & Molk.) M. Fleisch.; **Present status: Valid name**

Distribution in India: Western Himalayas, Eastern Himalayas and South India

931. *Floribundaria leptonema* (Müll. Hal.) Broth.; **Present status: Valid name**

Distribution in India: Western Himalayas and Eastern Himalayas

932. *Floribundaria sparsa* (Mitt.) Broth.; **Present status: Valid name**

Distribution in India: Eastern Himalayas and South India

933. *Floribundaria thuidioides* M. Fleisch.; **Present status: Valid name**

Distribution in India: South India

934. *Floribundaria walkeri* (Renauld & Cardot) Broth.; **Present status: Valid name**

Distribution in India: Western Himalayas, Eastern Himalayas and South India

176. *Meteoriopsis* M. Fleisch.

935. *Meteoriopsis ancistrodes* (Ren. & Card.) Broth. [Asthana & Sahu, 2013]; **Present status: Valid name**

Distribution in India: Western Himalayas

936. *Meteoriopsis divergens* (Mitt.) Broth.; Present status: **Synonym of** *Pterobryopsis divergens* (Mitt.) Nog.

Distribution in India: Eastern Himalayas

937. *Meteoriopsis formosana* Nog.; Present status: **Synonym of** *Meteoriopsis reclinata* (Müll. Hal.) M. Fleisch.

Distribution in India: Western Himalayas and Eastern Himalayas

938. *Meteoriopsis reclinata* (Müll. Hal.) M. Fleisch.; **Present status: Valid name**

Distribution in India: Western Himalayas, Eastern Himalayas, Central India and South India

939. *Meteoriopsis squarrosa* (Hook. & Harv.) M. Fleisch.; **Present status: Valid name**

Distribution in India: Western Himalayas, Eastern Himalayas and South India

177. *Meteorium* (Brid.) Dozy & Molk.

940. *Meteorium brevirameum* (Müll. Hal.) Broth.; Present status: **Synonym of** *Meteorium polytrichum* Dozy & Molk.

Distribution in India: South India

941. *Meteorium buchananii* (Brid.) Broth.; **Present status: Valid name**

Distribution in India: Western Himalayas, Eastern Himalayas and South India

942. *Meteorium helminthocladum* (Müll. Hal.) M. Fleisch.; Present status**: Synonym of** *Meteorium subpolytrichum* (Besch.) Broth.; Dozy & Molk.

Distribution in India: Eastern Himalayas

943. *Meteorium miquelianum* (Müll. Hal.) M. Fleisch.; Present status**: Synonym of** *Meteorium polytrichum* Dozy & Molk.

Distribution in India: South India

178. *Papillaria* (Müll. Hal.) Müll. Hal.

944. *Papillaria chrysoclada* (Müll. Hal.) A. Jaeger; **Present status: Valid name**

Distribution in India: Eastern Himalayas and South India

945. *Papillaria crocea* (Hampe) A. Jaeger; **Present status: Valid name**

Distribution in India: South India

946. *Papillaria feae* Müll. Hal. ex M. Fleisch.; **Present status: Valid name**

Distribution in India: South India

947. *Papillaria formosana* var. *pilifera* Nog. ; Present status**: Synonym of** *Trachycladiella sparsa* (Mitt.) M. Menzel

Distribution in India: Eastern Himalayas and South India

948. *Papillaria fuscescens* (Hook.) A. Jaeger; **Present status: Valid name**

Distribution in India: Eastern Himalayas and South India

949. *Papillaria semitorta* (Müll. Hal.) A. Jaeger ; **Present status: Valid name**

Distribution in India: Western Himalayas, Eastern Himalayas and South India

179. *Pseudospiridentopsis* (Broth.) M. Fleisch.

950. *Pseudospiridentosis horrida* (Cardot) Fleisch.; **Present status: Valid name**

Distribution in India: Eastern Himalayas

180. *Trachypodopsis* M. Fleisch.

951. *Trachypodopsis auriculata* (Mitt.) M. Fleisch.; **Present status: Valid name**

Distribution in India: Western Himalayas, Eastern Himalayas and South India

952. *Trachypodopsis serrulata* (P. Beauv.) M. Fleisch.; **Present status: Valid name**

Distribution in India: Western Himalayas, Eastern Himalayas, South India and Andaman Islands

181. *Trachypus* Reinw. & Hornsch.

953. *Trachypus bicolor* Reinw. & Hornsch.; **Present status: Valid name**

Distribution in India: Western Himalayas, Eastern Himalayas and South India

954. *Trachypus humilis* Lindb.; **Present status: Valid name**

Distribution in India: South India

Family: Entodontaceae Kindb.

182. *Campylodontium* Schwägr.

955. *Campylodontium flavescens* (Hook.) Bosch & Sande Lac.; **Present status: Valid name**

Distribution in India: Eastern Himalayas, South India and Gangetic Plains

956. *Campylodontium perplicatum* (Thér. & P. de la Varde) Broth. ; **Present status: Valid name**

Distribution in India: South India

183. *Erythrodontium* Hampe

957. *Erythrodontium julaceum* (Hook. & Schwägr.) Paris ; **Present status: Valid name**

Distribution in India: Western Himalayas, Eastern Himalayas, Central India and South India

184. *Orthotheciadelphus* Dixon

958. *Orthotheciadelphus ovicarpus* Dixon ; Present status: **Synonym of** *Orthothecium ovicarpum* (Dixon) W.R. Buck

Distribution in India: Western Himalayas

185. *Retidens* Dixon

959. *Retidens stewartii* Dixon ; *Present status*: **Synonym of** *Entodon stewartii* (Dixon) W.R. Buck

Distribution in India: Western Himalayas

Family: Plagiotheciaceae (Broth.) M. Fleisch.

186. *Plagiothecium* Bruch & Schimp.

960. *Plagiothecium cochleatum* Dixon ; Present status: **Doubtful name**

Distribution in India: Western Himalayas

961. *Plagiothecium dehradunense* Vohra; Present status: **Doubtful name**

Distribution in India: South India

962. *Plagiothecium denticulatum* (Hedw.) Schimp.; **Present status: Valid name**

Distribution in India: Western Himalayas, Eastern Himalayas and South India

963. *Plagiothecium entodontella* Broth. ex Dixon; Present status: **Doubtful name**

Distribution in India: Eastern Himalayas (Endemic to India)

964. *Plagiothecium neckeroideum* Schimp.; **Present status: Valid name**

Distribution in India: Western Himalayas, Eastern Himalayas and South India

965. *Plagiothecium neckeroideum* var. *sikkimense* Renauld & Cardot.; Present status: **Doubtful name**

Distribution in India: Western Himalayas, Eastern Himalayas and South India

966. *Plagiothecium nemorale* (Mitt.) A. Jaeger.; **Present status: Valid name**

Distribution in India: Western Himalayas and Eastern Himalayas

967. *Plagiothecium paleaceum* (Mitt.) A. Jaeger; **Present status: Valid name**

Distribution in India: Eastern Himalayas (Endemic to India)

968. *Plagiothecium perminutum* Dixon ; Present status: **Doubtful name**

Distribution in India: Western Himalayas

969. *Plagiothecium roseanum* Schimp.; **Present status: Valid name**

Distribution in India: Western Himalayas

970. *Plagiothecium sylvaticum* (Brid.) Schimp.; Present status**: Synonym of** *Plagiothecium denticulatum* subsp. *sylvaticum* (Brid.) Dixon

Distribution in India: Western Himalayas, Eastern Himalayas and Punjab

971. *Plagiothecium vesiculariopsis* Dixon & P. de la Varde; Present status: **Doubtful name**

Distribution in India: Eastern Himalayas

187. *Stereophyllum* Mitt.

972. *Stereophyllum acuminatum* Dixon & P. de la Varde; Present status: **Synonym of** *Entodontopsis leucostega* (Brid.) W.R. Buck & Ireland

Distribution in India: South India

973. *Stereophyllum anceps* (Bosch & Sande Lac.) Broth. ; Present status: **Synonym of** *Entodontopsis anceps* (Bosch & Sande Lac.) W.R. Buck & Ireland

Distribution in India: Eastern Himalayas, Central India and South India

974. *Stereophyllum confusion* Thér.; Present status: **Synonym of** *Stereophyllum radiculosum* (Hook.) Mitt.

Distribution in India: Central India and South India

975. *Stereophyllum decorum* (Mitt.) Wijk & Margad ; **Present status: Valid name**

Distribution in India: Western Himalayas

976. *Stereophyllum fulvum* (Harv.) A. Jaeger; **Present status: Valid name**

Distribution in India: Eastern Himalayas and South India

977. *Stereophyllum indicum* (Bél.) Mitt.; **Present status: Valid name**

Distribution in India: South India

978. *Stereophyllum ligulatum* Jaeger; Present status: **Synonym of** *Entodontopsis nitens* (Mitt.) W.R. Buck & R.R. Ireland

Distribution in India: South India and Central India

979. *Stereophyllum setschwanicum* Broth.; Present status: **Synonym of** *Entodontopsis setschwanica* (Broth.) W.R. Buck & Ireland

Distribution in India: Eastern Himalayas

980. *Stereophyllum subacuminatum* Dixon & P. de la Varde; **Present status: Doubtful**

Distribution in India: South India and Central India

981. *Stereophyllum tavoyense* (Hook. & Harv.) A. Jaeger.; **Present status: Valid name**

Distribution in India: Western Himalayas, South India and Gangetic plains

982. *Stereophyllum wightii* (Mitt.) A. Jaeger ; *Present status*: **Synonym of** *Entodontopsis wightii* (Mitt.) W.R. Buck & Ireland

Distribution in India: Western Himalayas, Eastern Himalayas, South India, Central India and Gangetic plains

Family: Regmatodontaceae Broth.

188. *Pseudobarbella* Nog.

983. *Pseudobarbella compressiramea* (Renauld & Cardot) Nog.; **Present status: Valid name**

Distribution in India: Eastern Himalayas

984. *Pseudobarbella levieri* (Renauld & Cardot) Nog.; **Present status: Valid name**

Distribution in India: Eastern Himalayas

189. *Regmatodon* Brid.

985. *Regmatodon declinatus* (Hook.) Brid.; **Present status: Valid name**

Distribution in India: Western Himalayas and Eastern Himalayas

986. *Regmatodon orthostegius* Mont.; **Present status: Valid name**

Distribution in India: Western Himalayas, Eastern Himalayas and South India

Family: Leucodontaceae Schimp.

190. ***Forsstroemia*** Lindb.

987. *Forsstroemia inclusa* Cardot & Dixon ; Present status: **Synonym of** *Forsstroemia thomsonii* (Mitt.) W.R. Buck

Distribution in India: Eastern Himalayas and Western Himalayas

988. *Forsstroemia indica* (Mont.) Paris; **Present status: Valid name**

Distribution in India: India Orientalis

989. *Forsstroemia secunda* Dixon & Badhw.; *Present status*: **Synonym of** ***Herpetineuron*** *acutifolium* (Mitt.) Granzow

Distribution in India: Western Himalayas

191. ***Leucodon*** Schwägr.

990. *Leucodon sciuroides* (Hedw.) Schwägr.; **Present status: Valid name**

Distribution in India: Western Himalayas

991. *Leucodon secundus* (Harv.) Mitt.; **Present status: Valid name**

Distribution in India: Eastern Himalayas, Western Himalayas and South India

Family: Amblystegiaceae G. Roth.

192. ***Amblystegium*** Bruch & Schimp.

992. *Amblystegium juratzkanum* Schimp.; Present status: **Synonym of** *Amblystegium serpens* (Hedw.) Schimp.

Distribution in India: Western Himalayas

993. *Amblystegium saxatile* Schimp.; *Present status*: **Synonym of** *Campylium radicale* (P. Beauv.) Grout

Distribution in India: Western Himalayas

994. *Amblystegium serpens* (Hedw.) Schimp.; **Present status: Valid name**

Distribution in India: Western Himalayas

995. *Amblystegium serpens* var. *saxicola* (Hesselbo) C.E.O. Jensen ; **Present status: Valid name**

Distribution in India: Western Himalayas

996. *Amblystegium sparsile* (Mitt.) Paris; **Present status: Valid name**

Distribution in India: Eastern Himalayas

997. *Amblystegium tibetanum* (Mitt.) Paris; **Present status: Valid name**

Distribution in India: Western Himalayas

998. *Amblystegium varium* (Hedw.) Lindb.; **Present status: Valid name**

Distribution in India: Western Himalayas

193. *Hygroamblystegium* Loeske

999. *Hygroamblystegium gangulianum* Vohra ; Present status: **Synonym of** *Brachythecium gangulianum* (Vohra) Ochyra

Distribution in India: Western Himalayas (Endemic to India)

1000. *Hygroamblystegium obtusulum* (Mitt.) Broth.; **Present status: Valid name**

Distribution in India: Western Himalayas

1001. *Hygroamblystegium tenax* (Hedw.) Jenn.; **Present status: Valid name**

Distribution in India: Western Himalayas

194. *Sciaromium* (Mitt.) Mitt.

1002. *Sciaromium sikkimense* Paris; **Present status: Synonym of** *Handeliobryum sikkimense* (Paris) Ochyra

Distribution in India: Eastern Himalayas (Endemic to India)

Family: Campyliaceae (Kanda) W. R. Buck

195. *Campyliadelphus* (Kindb.) R. S. Chopra

1003. *Campyliadelphus elodes* (Lindb.) Kanda ; **Present status: Valid name**

Distribution in India: Western Himalayas

1004. *Campyliadelphus protensus* (Brid.) Kanda ; Present status: **Synonym of** *Campyliadelphus stellatus* var. *protensus* (Brid.) Ochyra

Distribution in India: Western Himalayas

196. *Campylium* (Sull.) Mitt.

1005. *Campylium chrysophyllum* (Brid.) J. Lange ; **Present status: Valid name**

Distribution in India: Western Himalayas

1006. *Campylium gollanii* Müll. Hal.; Cuurent status: **Doubtful name**

Distribution in India: Western Himalayas (Endemic to India)

1007. *Campylium lacerulum* (Mitt.) Broth.; **Present status: Valid name**

Distribution in India: Eastern Himalayas

1008. *Campylium halleri* (Sw. & Hedw.) Lindb.; **Present status: Valid name**

Distribution in India: Western Himalayas and Eastern Himalayas

1009. *Campylium sommerfeltii* (Myrin) J. Lange ; **Present status: Valid name**

Distribution in India: Western Himalayas

Family: Cratoneuraceae Mönk.

197. *Cratoneuron* (Sull.) Spruce

1010. *Cratoneuron commutatum* (Hedw.) G. Roth.; **Present status: Valid name**

Distribution in India: Western Himalayas

1011. *Cratoneuron commutatum* var. *falcatum* (Brid.) Mönk.**; Present status: Valid name**

Distribution in India: Western Himalayas

1012. *Cratoneuron filicinum* (Hedw.) Spruce ; **Present status: Valid name**

Distribution in India: Western Himalayas

1013. *Cratoneuron filicinum* var. *fallax* (Brid.) G. Roth.; **Present status: Valid name**

Distribution in India: Western Himalayas

198. ***Drepanocladus*** (Müll. Hal.) G. Roth.

1014. *Drepanocladus aduncus* (Hedw.) Warnst.; **Present status: Valid name**

Distribution in India: Western Himalayas and Eastern Himalayas

1015. *Drepanocladus exannulatus* (Schimp.) Warnst.; **Present status: Valid name**

Distribution in India: Western Himalayas

1016. *Drepanocladus fluitans* (Hedw.) Warnst.; **Present status: Valid name**

Distribution in India: Western Himalayas

1017. *Drepanocladus uncinatus* (Hedw.) Warnst.; Present status**: Synonym of** *Sanionia uncinata* (Hedw.) Loeske

Distribution in India: Western Himalayas and Eastern Himalayas

199. ***Hygrohypnum*** Lindb.

1018. *Hygrohypnum choprae* Vohra; **Present status: Valid name**

Distribution in India: Eastern Himalayas

1019. *Hygrohypnum dilatatum* (Wilson) Loeske ; *Present status*: **Synonym of** *Hygrohypnum duriusculum* (De Not.) D.W. Jamieson

Distribution in India: Western Himalayas

1020. *Hygrohypnum luridum* (Hedw.) Jenn.; **Present status: Valid name**

Distribution in India: Western Himalayas

1021. *Hygrohypnum nairii* Vohra ; Present status: **Doubtful name Endemic to India**

Distribution in India: Western Himalayas

200. ***Leptodictyum*** (Schunp.) Warnst.

1022. *Leptodictyum riparium* (Hedw.) Warnst.; **Present status: Valid name**

Distribution in India: Western Himalayas

Family: Neckeraceae Schimp.

201. ***Cryptoleptodon*** Renauld & Cardot

1023. *Cryptoleptodon flexuosus* (Harv.) Renauld & Cardot ; **Present status: Valid name**

Distribution in India: Western Himalayas and Eastern Himalayas

1024. *Cryptoleptodon pluvinii* (Brid.) Broth.; **Present status: Valid name**

Distribution in India: Western Himalayas

1025. *Cryptoleptodon rigidulus* (Wilson & Mitt.) Broth.**; Present status: Valid name**

Distribution in India: Western Himalayas

202. *Handeliobryum* Broth.

1026. *Handeliobryum assamicum* Dixon; *Present status*: **Synonym of** *Handeliobryum sikkimense* (Paris) Ochyra

Distribution in India: Eastern Himalayas

1027. *Handeliobryum himalayanum* Broth.; *Present status*: **Synonym of** *Handeliobryum sikkimense* (Paris) Ochyra

Distribution in India: Eastern Himalayas

1028. *Handeliobryum setschwanicum* Broth.; *Present status*: **Synonym of** *Handeliobryum sikkimense* (Paris) Ochyra

Distribution in India: Eastern Himalayas and South India

203. *Himantocladium* (Mitt.) Fleisch.

1029. *Himantocladium cyclophyllum* (Müll. Hal.) M. Fleisch.; **Present status: Valid name**

Distribution in India: Eastern Himalayas and South India

1030. *Himantocladium flagelliferum* (Broth.) Broth.; **Present status: Valid name**

Distribution in India: Eastern Himalayas

1031. *Himantocladium loriforme* (Bosch & Sande Lac.) M. Fleisch.; **Present status: Valid name**

Distribution in India: Eastern Himalayas

1032. *Himantocladium plumula* (Nees) M. Fleisch.; **Present status: Valid name**

Distribution in India: Central India

1033. *Himantocladium rugulosum* (Mitt.) M. Fleisch.; **Present status: Valid name**

Distribution in India: South India

1034. *Himantocladium strictum* Dixon; *Present status*: **Synonym of** *Neckera crenulata* Harv.

Distribution in India: South India

204. *Homalia* (Brid.) Bruch & Schimp.

1035. *Homalia obtusata* Mitt.; *Present status*: **Synonym of** *Homalia trichomanoides* (Hedw.) Schimp.

Distribution in India: Western Himalayas

1036. *Homalia pygmaea* Broth.; **Present status: Valid name**

Distribution in India: South India

205. *Homaliodendron* Fleisch

1037. *Homaliodendron dentatum* (Griff.) M. Fleisch.; **Present status: Valid name**

Distribution in India: Eastern Himalayas

1038. *Homaliodendron exiguum* (Bosch & Lac.) M. Fleisch. [Asthana & Sahu, 2013]; **Present status: Valid name**

Distribution in India: Western Himalayas, Eastern Himalayas and South India

1039. *Homaliodendron flabellatum* (Sm.) M. Fleisch.; **Present status: Valid name**

Distribution in India: Eastern Himalayas

1040. *Homaliodendron javanicum* (Müll. Hal.) M. Fleisch.; **Present status: Valid name**

Distribution in India: Central India

1041. *Homaliodendron ligulifolium* (Mitt.) M. Fleisch.; **Present status: Valid name**

Distribution in India: Eastern Himalayas

1042. *Homaliodendron microdendron* (Mont.) M. Fleisch.; **Present status: Valid name**

Distribution in India: Western Himalayas, Eastern Himalayas and South India

1043. *Homaliodendron montagneanum* (Müll. Hal.) M. Fleisch.; **Present status: Valid name**

Distribution in India: Eastern Himalayas and South India

1044. *Homaliodendron obtusatum* (Mitt.) Gangulee ; *Present status*: **Synonym of** *Homalia trichomanoides* (Hedw.) Schimp.

Distribution in India: Western Himalayas and Eastern Himalayas

1045. *Homaliodendron paquei* (Renauld & Cardot) Broth.; **Present status: Valid name**

Distribution in India: Eastern Himalayas

1046. *Homaliodendron rectifolium* (Mitt.) M. Fleisch.; **Present status: Valid name**

Distribution in India: Eastern Himalayas

1047. *Homaliodendron scalpellifolium* (Mitt.) M. Fleisch.; Present **status: Valid name**

Distribution in India: Western Himalayas and Eastern Himalayas

1048. *Homaliodendron sphaerocarpum* Nog.; **Present status: Valid name**

Distribution in India: Eastern Himalayas

1049. *Homaliodendron stracheyanum* (Mitt.) M. Fleisch. ; **Present status: Valid name**

Distribution in India: Western Himalayas

206. *Neckera* Hedw.

1050. *Neckera aequalifolia* Müll. Hal.; Present status: **Doubtful name**

Distribution in india: South India

1051. *Neckera andrei* Thér. & P. de la Varde; Present status: **Doubtful name**

Distribution in india: South India

1052. *Neckera complanata* (Hedw.) Huebener ; **Present status: Valid name**

Distribution in India: Western Himalayas

1053. *Neckera crenulata* Harv.; **Present status: Valid name**

Distribution in India: Western Himalayas and Eastern Himalayas

1054. *Neckera foreaui* Harv.;.; **Present status: Doubtful name**

Distribution in India: South India

1055. *Neckera goughiana* Mitt.; **Present status: Valid name**

Distribution in india: South India

1056. *Neckera himalayana* Mitt.; **Present status: Valid name**

Distribution in india: Eastern and Western Himalayas

1057. *Neckera pennata* Hedw.; **Present status: Valid name**

Distribution in India: Western Himalayas, Eastern Himalayas and South India

1058. *Neckera semicrispa* Cardot & P. de la Varde; Present status: **Doubtful name**

Distribution in india: South India

1059. *Neckera setschwanica* Broth.; Present status: **Doubtful name**

Distribution in india: Eastern Himalayas

207. *Neckeropsis* Reichdt.

1060. *Neckeropsis acutata* (Mitt.) M. Fleisch.; **Present status: Valid name**

Distribution in india: Eastern Himalayas

1061. *Neckeropsis andamana* (Müll. Hal.) M. Fleisch.; **Present status: Valid name**

Distribution in india: South India, Andaman and Nicobar Islands

1062. *Neckeropsis crinita* (Griff.) M. Fleisch.; **Present status: Valid name**

Distribution in india: Eastern Himalayas and Central India

1063. *Neckeropsis darjeelingensis* Gangulee; Present status: **Doubtful name**

Distribution in india: Eastern Himalayas (Endemic to India)

1064. *Neckeropsis exserta* (Hook. & Schwägr.) Broth.; **Present status: Valid name**

Distribution in India: Western Himalayas, Eastern Himalayas and South India

1065. *Neckeropsis fimbriata* (Harv.) M. Fleisch.; Present status: **Valid name**

Distribution in india: Eastern Himalayas

1066. *Neckeropsis gracilenta* (Bosch & Sande Lac.) M. Fleisch.; Present status: **Valid name**

Distribution in india: Eastern Himalayas and Nicobar Islands

1067. *Neckeropsis lepineana* (Mont.) M. Fleisch.; **Present status: Valid name**

Distribution in India: Western Himalayas, Eastern Himalayas and South India

1068. *Neckeropsis submarginata* Cardot ex Touw; Present status: **Valid name**

Distribution in india: South India

208. *Pinnatella* M. Fleisch.

1069. *Pinnatella alopecuroides* (Mitt.) M. Fleisch.; **Present status: Valid name**

Distribution in India: Eastern Himalayas and South India

1070. *Pinnatella ambigua* (Bosch & Sande Lac.) M. Fleisch.; **Present status: Valid name**

Distribution in India: Western Himalayas, Eastern Himalayas and Central India

1071. *Pinnatella anacamptolepis* (Müll. Hal.) Broth.; **Present status: Valid name**

Distribution in India: South India

1072. *Pinnatella calcutensis* M. Fleisch.; **Present status: Valid name**

Distribution in India: South India, Central India and Gangetic plains

1073. *Pinnatella foreauana* Thér. & P. de la Varde; **Present status: Valid name**

Distribution in India: South India

1074. *Pinnatella gollanii* Broth.; Present status: **Doubtful name**

Distribution in India: Western Himalayas

1075. *Pinnatella kurzii* (Kindb.) Wijk & Margad.; Present status: **Synonym of** *Curvicladium kurzii* (Kindb.) Enroth

Distribution in India: Eastern Himalayas

1076. *Pinnatella limbata* Dixon; Present status: **Doubtful name**

Distribution in India: South India

1077. *Pinnatella microptera* M. Fleisch.; Present status: **Synonym of** *Caduciella mariei* (Besch.) Enroth

Distribution in India: Eastern Himalayas

1078. *Pinnatella sikkimensis* Broth. ; Present status: **Synonym of** *Pinnatella foreauana* Thér. & P. de la Varde

Distribution in India: Eastern Himalayas

208. *Porotrichum* (Brid.) Hampe

1079. *Porotrichum fruticosum* Mitt.; Present status: **Synonym of** *Camptochaete fruticosa* Paris

Distribution in India: Western Himalayas and South India

209. *Thamnobryum* Nieuwl.

1080. *Thamnobryum alopecurum* (Hedw.) Nieuwl. & Gangulee ; **Present status: Valid name**

Distribution in India: Western Himalayas and South India

1081. *Thamnobryum fruticosum* (Mitt.) Gangulee; **Present status:** Synonym of *Homaliodendron fruticosum* (Mitt.) S. Olsson, Enroth & D. Quandt

Distribution in India: Eastern Himalayas

1082. *Thamnobryum latifolium* (Bosch & Sande Lac.) Nieuwl.; **Present status: Valid name**

Distribution in India: Western Himalayas

1083. *Thamnobryum parvulum* (Mitt.) R.S. Chopra; **Present status: Valid name**

Distribution in India: Western Himalayas and South India

1084. *Thamnobryum schmidii* (Müll. Hal.) R.S. Chopra; **Present status: Valid name**

Distribution in India: South India

1085. *Thamnobryum subseriatum* (Mitt. & Sande Lac.) B.C. Tan ; **Present status: Valid name**

Distribution in India: Western Himalayas, Eastern Himalayas and South India

Family: Hylocomiaceae (Broth.) M. Fleisch.

210. *Cyathothecium* Dixon

1086. *Cyathothecium distichaceum* Dixon ; Present status: **Doubtful name**

Distribution in India: Western Himalayas

211. *Hylocomium* Bruch & Schimp.

1087. *Hylocomium himalayanum* (Mitt.) A. Jaeger ; Present status: **Synonym of** *Hylocomiastrum himalayanum* (Mitt.) Broth.

Distribution in India: Western Himalayas and Eastern Himalayas

1088. *Hylocomium indicum* Dixon ; *Present status*: **Synonym of** *Pseudopleuropus indicus* (Dixon) T.Y. Chiang

Distribution in India: Western Himalayas

212. *Leptohymenium* Schwägr.

1089. *Leptohymenium tenue* (Hook.) Schwägr.; **Present status: Valid name**

Distribution in India: Western Himalayas and Eastern Himalayas

213. *Macrothamnium* M. Fleisch.

1090. *Macrothamnium macrocarpum* (Reinw. & Hornsch.) M. Fleisch.; **Present status: Valid name**

Distribution in India: Western Himalayas, Eastern Himalayas and South India

1091. *Macrothamnium pseudostriatum* (Müll. Hal.) M. Fleisch; Present status: **Synonym** of *Macrothamnium macrocarpum* (Reinw. & Hornsch.) M. Fleisch.

Distribution in India: Eastern Himalayas

1092. *Macrothamnium stigmatophyllum* M. Fleisch.; Present status: **Synonym of** *Macrothamnium submacrocarpum* (A. Jaeger ex Renauld & Cardot) M. Fleisch.

Distribution in India: Eastern Himalayas

1093. *Macrothamnium submacrocarpon* A. Jaeger et Renauld ex Cardot ; Present status: **Synonym of** *Macrothamnium macrocarpum* (Reinw. & Hornsch.) M. Fleisch.

Distribution in India: Western Himalayas, Eastern Himalayas and South India

214. *Meteoriella* S. Okamura

1094. *Meteoriella soluta* (Mitt.) S. Okamura; **Present status: Valid name**

Distribution in India: Eastern Himalayas

215. *Neodolichomitra* Nog.

1095. *Neodolichomitra robusta* (Broth.) Nog.; Present status: **Synonym of** *Neodolichomitra yunnanensis* (Besch.) T.J. Kop.

Distribution in India: Eastern Himalayas

216. *Orontobryum* M. Fleisch.

1096. *Orontobryum hookeri* (Mitt.) M. Fleisch.; **Present status: Valid name**

Distribution in India: Eastern Himalayas

1097. *Orontobryum hookeri* (Mitt.) M. Fleisch.; **Present status: Valid name**

Distribution in India: Eastern Himalayas

217. *Pleurozium* Mitt.

1098. *Pleurozium schreberi* (Willd. ex Brid.) Mitt.; **Present status: Valid name**

Distribution in India: Eastern Himalayas

218. *Rhytidiadelphus* (Limpr.) Warnst.

1099. *Rhytidiadelphus triquetrus* (Hedw.) Warnst.; **Present status: Valid name**

Distribution in India: Western Himalayas and Eastern Himalayas

219. *Stenotheciopsis* M. Fleisch.

1100. *Stenotheciopsis serrula* Mitt.; Present status: **Synonym of** *Stenotheciopsis serrula* (Mitt.) M. Fleisch.

Distribution in India: Western Himalayas and Eastern Himalayas

Family: Leskeaceae Schimp.

220. *Anomodon* Hook. & Taylor

1101. *Anomodon acutifolius* Mitt.; **Present status: Valid name**

Distribution in India: Western Himalayas and Eastern Himalayas

1102. *Anomodon attenuatus* (Hedw.) Huebener; **Present status: Valid name**

Distribution in India: Western Himalayas

1103. *Anomodon minor* (Hedw.) Fürnr.; **Present status: Valid name**

Distribution in India: Western Himalayas and Eastern Himalayas

1104. *Anomodon minor* (Hedw.) Lindb. ssp. *integerrimus* (Mitt.) Z. Iwats.; **Present status: Valid name**

Distribution in India: Western Himalayas and Eastern Himalayas

1105. *Anomodon planatus* Mitt.; Present status: **Synonym of** *Anomodon minor* (Hedw.) Lindb.

Distribution in India: Western Himalayas and Eastern Himalayas

1106. *Anomodon rostratus* (Hedw.) Schimp.; **Present status: Valid name**

Distribution in India: Western Himalayas

1107. *Anomodon rugelii* (Müll. Hal.) Keissl.; **Present status: Valid name**

Distribution in India: Western Himalayas and Eastern Himalayas

1108. *Anomodon thrautus* Müll. Hal.; **Present status: Valid name**

Distribution in India: Western Himalayas

1109. *Anomodon viticulosus* (Hedw.) Hook. & Taylor ; **Present status: Valid name**

Distribution in India: Western Himalayas and Eastern Himalayas

221. ***Habrodon*** Schimp.

1110. *Habrodon kashmiriensis* Vohra ; Present status: **Doubtful name**

Distribution in India: Western Himalayas (Endemic to India)

222. ***Haplohymenium*** Dozy & Molk.

1111. *Haplohymenium triste* (Ces.) Kindb.; **Present status: Valid name**

Distribution in India: Western Himalayas

223. ***Herpetineuron*** (Müll. Hal.) Cardot

1112. *Herpetineuron toccoae* (Sull. & Lesq.) Cardot ; **Present status: Valid name**

Distribution in India: Western Himalayas, Eastern Himalayas, South India and Central India

224. ***Leptopterigynandrum*** Müll. Hal.

1113. *Leptopterigynandrum autoicum* Dixon ex Gangulee et Vohra; Present status: **Doubtful name**

Distribution in India: Indian orientalis

1114. *Leptopterigynandrum decolor* (Mitt.) Fleisch.; **Present status: Valid name**

Distribution in India: Eastern Himalayas (Endemic to India)

1115. *Leptopterigynandrum brevirete* Dixon ; Present status: **Doubtful name**

Distribution in India: Western Himalayas

1116. *Leptopterigynandrum subintegrum* (Mitt.) Broth.; **Present status: Valid name**

Distribution in India: Western Himalayas

225. ***Lescuraea*** Bruch & Schimp.

1117. *Lescuraea darjeelingensis* Vohra; Present status: **Doubtful name**

Distribution in India: Eastern Himalayas

1118. *Lescuraea incurvata* (Hedw.) E. Lawton ; **Present status: Valid name**

Distribution in India: Western Himalayas

1119. *Lescuraea laevifolia* (Mitt.) R. S. Chopra ; **Present status: Valid name**

Distribution in India: Western Himalayas

1120. *Lescuraea mutabilis* (Brid.) Lindb.; **Present status: Valid name**

Distribution in India: Western Himalayas

1121. *Lescuraea ramuligera* (Mitt.) R. S. Chopra ; **Present status: Valid name**

Distribution in India: Western Himalayas

1122. *Lescuraea saxicola* (Bruch & Schimp.) Molendo ; **Present status: Valid name**

Distribution in India: Western Himalayas

226. *Leskea* Hedw.

1123. *Leskea consanguinea* (Mont.) Mitt.; Present status**: Synonym of** *Leskeella consanguinea* (Mont.) Broth.

Distribution in India: South India

1124. *Leskea hyalopiculata* Dixon ; Present status: **Doubtful name**

Distribution in India: Western Himalayas

1125. *Leskea perstricta* Dixon; Present status: **Doubtful name**

Distribution in India: Eastern Himalayas (Endemic to India)

227. *Leskeella* (Limpr.) Loesk.

1126. *Leskeella incrassata* (Lindb. & Broth.) Broth.; **Present status: Valid name**

Distribution in India: Western Himalayas

1127. *Leskeella nervosa* (Brid.) Loesk.; **Present status: Valid name**

Distribution in India: Western Himalayas, Punjab Plains

228. *Lindbergia* Kindb.

1128. *Lindbergia duthiei* (Broth.) Broth.; **Present status: Valid name**

Distribution in India: Western Himalayas

1129. *Lindbergia koelzii* R. S. Williams ; Present status: **Doubtful name**

Distribution in India: Western Himalayas

1130. *Lindbergia longinervis* Cardot. & Dixon ; Present status: **Doubtful name**

Distribution in India: Western Himalayas

229. *Pseudoleskea* Bruch & Schimp.

1131. *Pseudoleskea incurvata* (Hedw.) Loeske; **Present status: Valid name**

Distribution in India: Eastern Himalayas

1132. *Pseudoleskea laevifolia* (Mitt.) A. Jaeger.; Present status**: Synonym of** *Lescuraea laevifolia* (Mitt.) R.S. Chopra

Distribution in India: Western Himalayas and South India

1133. *Pseudoleskea ramuligera* (Mitt.) Sauerb. & A. Jaeger; Present status**: Synonym of** *Lescuraea ramuligera* (Mitt.) R.S. Chopra

Distribution in India: Eastern Himalayas

230. *Pseudoleskeella* Kindb.

1134. *Pseudoleskeella catenulata* (Brid. & Schrad.) Kindb.; **Present status: Valid name**

Distribution in India: Western Himalayas

231. *Pseudoleskeopsis* Broth.

1135. *Pseudoleskeopsis decurvata* (Mitt.) Broth. & Dixon; **Present status: Valid name**

Distribution in India: Eastern Himalayas and Western Himalayas

1136. *Pseudoleskeopsis orbiculata* (Mitt.) Broth.; **Present status: Valid name**

Distribution in India: Eastern Himalayas and South India

1137. *Pseudoleskeopsis serrulata* Cardot & Thér.; Present status**: Synonym of** *Pseudoleskeopsis zippelii* (Dozy & Molk.) Broth.

Distribution in India: Western Himalayas

1138. *Pseudoleskeopsis zippelii* (Dozy & Molk.) Broth.; **Present status: Valid name**

Distribution in India: Eastern Himalayas, Western Himalayas and South India

232. ***Okamuraea*** Broth.

1139. *Okamuraea hakoniensis* (Mitt.) Broth.; **Present status: Valid name**

Distribution in India: India Orientalis

233. ***Schwetschkea*** C. Muell.

1140. *Schwetschkea applanata* (Thwaites & Mitt.) Broth.; **Present status: Valid name**

Distribution in India: South India

1141. *Schwetschkea indica* Broth.; **Present status:** Synonym of *Macgregorella indica* (Broth.) W.R. Buck

Distribution in India: South India

Family: Sematophyllaceae Broth.

234. ***Acroporium*** Mitt.

1142. *Acroporium affine* Broth.; **Present status: Valid name**

Distribution in India: South India

1143. *Acroporium baviense* (Besch.) Broth.; **Present status: Valid name**

Distribution in India: Eastern Himalayas

1144. *Acroporium vincensianum* (Thér.) Broth.; **Present status: Valid name**

Distribution in India: South India

235. ***Aptychella*** (Broth.) Herzog

1145. *Aptychella borii* Dixon; Present status**: Synonym of** *Clastobryopsis planula* var. *delicata* (M. Fleisch.) B.C. Tan & Y. Jia

Distribution in India: Eastern Himalays (Endemic to India)

1146. *Aptychella delicata* (M. Fleisch.) M. Fleisch.; **Present status: Valid name**

Distribution in India: Eastern Himalays

1147. *Aptychella muelleri* Dixon; Present status: **Synonym of** *Clastobryopsis muelleri* (Dixon) Tixier

Distribution in India: Eastern Himalays

1148. *Aptychella planula* (Mitt.) M. Fleisch.; Present status: **Synonym of** *Clastobryum planulum* (Mitt.) Brühl

Distribution in India: Eastern Himalays

1149. *Aptychella serrulata* (Cardot & P. de la Varde) Broth.; **Present status: Valid name**

Distribution in India: South India

1150. *Aptychella tenuiramea* (Mitt.) Tixier ; **Present status: Valid name**

Distribution in India: Eastern Himalayas, Western Himalayas and South India

236. *Brotherella* Loesk. & M. Fleisch.

1151. *Brotherella amblystega* (Mitt.) Broth.; Present status: **Synonym of** *Pylaisiadelpha amblystega* (Mitt.) W.R. Buck

Distribution in India: Eastern Himalyas (Endemic to India)

1152. *Brotherella curvirostris* (Schwägr.) M. Fleisch.; **Present status: Valid name**

Distribution in India: Eastern Himalyas

1153. *Brotherella dixonii* Herzog; Present status: **Synonym of** *Ectropothecium dixonii* (Herzog) Y. Jia & S. He

Distribution in India: Eastern Himalyas (Endemic to India)

1154. *Brotherella erythrocaulis* (Mitt.) M. Fleisch.; Present status: **Synonym of** *Pylaisiadelpha erythrocaulis* (Mitt.) W.R. Buck

Distribution in India: Eastern Himalyas

1155. *Brotherella falcata* (Dozy & Molk.) M. Fleisch.; **Present status: Valid name**

Distribution in India: Eastern Himalyas

1156. *Brotherella filiformis* Dixon; Present status: **Synonym of** *Pylaisiadelpha filiformis* (Dixon) W.R. Buck

Distribution in India: Eastern Himalyas (Endemic to India)

1157. *Brotherella harveyana* (Mitt.) Dixon ; **Present status: Valid name**

Distribution in India: Western Himalyas

1158. *Brotherella nictans* (Mitt.) Broth.; **Present status: Valid name**

Distribution in India: Eastern Himalyas

1159. *Brotherella pallida* (Renauld & Cardot) M. Fleisch.; **Present status: Valid name**

Distribution in India: Eastern Himalyas(Endemic to India)

1160. *Brotherella perpinnata* (Broth.) M. Fleisch.; Present status: **Synonym of** *Pylaisiadelpha perpinnata* (Broth.) W.R. Buck

Distribution in India: Western Himalyas and Eastern Himalyas

1161. *Brotherella propinqua* (Harv.) M. Fleisch.; **Present status: Valid name**

Distribution in India: Eastern Himalyas

1162. *Brotherella yokahamae* (Broth.) Broth.; **Present status: Valid name**

Distribution in India: Western Himalyas and Eastern Himalyas

237. *Chinostomum* Müll. Hal.

1163. *Chinostomum rostratum* (Griff.) C. Muell.; **Present status: Valid name**

Distribution in India: Eastern Himalayas and South India

238. *Clastobryella* M. Fleisch.

1164. *Clastobryella gracilis* Dixon & P. de la Varde; **Present status:** Synonym of *Gammiella capillacea* (Griff.) Tixier

Distribution in India: South India

239. *Clastobryopsis* M. Fleisch.

1165. *Clastobryopsis muelleri* (Dixon) Tixier; **Present status: Valid name**

Distribution in India: Eastern Himalayas

1166. *Clastobryopsis planula* (Mitt.) M. Fleisch.; **Present status: Valid name**

Distribution in India: Eastern Himalayas

240. *Clastobryum* Dozy & Molk.

1167. *Clastobryum barbelloides* Dixon & P. de la Varde; **Present status:** Synonym of *Aptychella tenuiramea* (Mitt.) Tixier

Distribution in India: South India

1168. *Clastobryum capillaceum* (Griff.) Broth.; Present status: **Synonym of** *Pylaisiadelpha capillacea* (Griff.) B.C. Tan & Y. Jia

Distribution in India: Eastern Himalayas, Western Himalayas and South India

1169. *Clastobryum cupressinoides* Dixon & P. de la Varde; **Present status:** Synonym of *Gammiella capillacea* (Griff.) Tixier

Distribution in India: South India

1170. *Clastobryum patentifolium* Dixon & P. de la Varde; **Present status:** Synonym of *Gammiella capillacea* (Griff.) Tixier

Distribution in India: South India

1171. *Clastobryum surculare* Dixon; **Present status: Valid name**

Distribution in India: Eastern Himalayas (Endemic to India)

1172. *Clastobryum wichurae* Dixon; **Present status: Valid name**

Distribution in India: Eastern Himalayas (Endemic to India)

241. *Heterophyllium* (Scxhimp.) Kindb.

1173. *Heterophyllium confine* (Mitt.) M. Fleisch.; Present Status: Synonym of *Heterophyllium affine* (Hook.) M. Fleisch.

Distribution in India: India Orientalis

1174. *Heterophyllium haldanianum* (Grev.) M. Fleisch.; **Present status: Valid name**

Distribution in India: Western Himalayas

1175. *Heterophyllium renitens* (Mitt.) Broth.; Present Status: **Synonym of** *Herzogiella renitens* (Mitt.) Z. Iwats.

Distribution in India: Eastern Himalayas

242. *Meiothecium* Mitt.

1176. *Meiothecium jagorii* (Müll. Hal.) Broth.; **Present status: Valid name**

Distribution in India: South India and Andaman Islands

1177. *Meiothecium microcarpum* (Harv.) Mitt.; **Present status: Valid name**

Distribution in India: South India and Andaman Islands

243. *Pylaisiadelpha* Cardot

1178. *Pylaisiadelpha drepanioides* Cardot. & Dixon ; Present status: **Doubtful name**

Distribution in India: Eastern Himalayas and Western Himalayas

244. *Pylaisiopsis* Broth.

1179. *Pylaisiopsis speciosa* (Mitt.) Broth.; Present Status: Synonym of *Aptychella speciosa* (Mitt.) Tixier

Distribution in India: Eastern Himalayas (Endemic to India)

245. *Rhaphidorrhynchium* M. Fleisch.

1180. *Rhaphidorrhynchium confertissimum* (Mitt.) Broth.; **Present status: Valid name**

Distribution in India: Eastern Himalayas (Endemic to India)

246. *Rhaphidostichum* M. Fleisch.

1181. *Rhaphidostichum complanatum* (Dixon) Dixon; **Present status:** Synonym of *Papillidiopsis complanata* (Dixon) W.R. Buck & B.C. Tan

Distribution in India: South India

1182. *Rhaphidostichum cucullifolium* (Cardot & Dixon) Broth. .; **Present status: Valid name**

Distribution in India: South India

1183. *Rhaphidostichum glauco-virens* (Mitt.) Broth. .; **Present status: Valid name**

Distribution in India: Eastern Himalayas

1184. *Rhaphidostichum luxurians* (Dozy & Molk.) M. Fleisch.; **Present status:** Synonym of *Papillidiopsis luxurians* (Dozy & Molk.) W.R. Buck & B.C. Tan

Distribution in India: Eastern Himalayas

1185. *Rhaphidostichum subleptocarpum* (Thér. & P. de la Varde) Broth.; **Present status: Valid name**

Distribution in India: South India

246. *Sematophyllum* Mitt.

1186. *Sematophyllum angusticuspis* Broth.; **Present status: Doubtful**

Distribution in India: South India

1187. *Sematophyllum caespitosum* Mitt.; **Present status:** Synonym of *Sematophyllum subpinnatum* (Brid.) E. Britton

Distribution in India: Eastern Himalayas and South India

1188.*Sematophyllum humile* (Mitt.) Broth.; **Present status: Valid name**

Distribution in India: Western Himalayas

1189. *Sematophyllum micans* (Mitt.) Braithw.; **Present status: Valid name**

Distribution in India: Eastern Himalayas

1190. *Sematophyllum phoeniceum* (Müll. Hal.) M. Fleisch.; **Present status: Valid name**

Distribution in India: South India, Anadaman and Nicobar Islands

1191. *Sematophyllum sebillei* (Broth. & Thér.) R.S. Chopra; **Present status: Valid name**

Distribution in India: South India

1192. *Sematophyllum subcylindricum* (Broth. ex M. Fleisch.) Sainsbury; **Present status: Valid name**

Distribution in India: South India

1193. *Sematophyllum subhumile* (Müll. Hal.) M. Fleisch.; **Present status: Valid name**

Distribution in India: Eastern Himalayas and South India

247. *Struckia* Müll. Hal.

1194. *Struckia argentata* (Mitt.) Müll. Hal.; **Present status: Valid name**

Distribution in India: Western Himalayas and Eastern Himalayas

1195. *Struckia griffithii* Müll. Hal.; **Present status:** Synonym of *Bryum argenteum* var. *griffithii* (Müll. Hal.) Gangulee

Distribution in India: Eastern Himalayas

248. *Taxithelium* Spruce *ex* Mitt.

1196. *Taxithelium instratum* (Brid.) Broth.; **Present status: Valid name**

Distribution in India: South India

1197. *Taxithelium kerianum* (Broth.) Broth.; **Present status: Valid name**

Distribution in India: Eastern Himalayas and Nicobar Islands

1198. *Taxithelium laeviusculum* Dixon; **Present status: Doubtful**

Distribution in India: Eastern Himalayas (Endemic to India)

1199. *Taxithelium nepalense* (Schwägr.) Broth.; **Present status: Valid name**

Distribution in India: Eastern Himalayas, South India, Central India, Andaman and Nicobar Islands

1200. *Taxithelium vernieri* (Duby) Besch.; **Present status: Valid name**

Distribution in India: Andaman and Nicobar Islands

249. *Trichosteleum* Mitt.

1201. *Trichosteleum boschii* (Dozy & Molk.) A. Jaeger; **Present status: Valid name**

Distribution in India: Eastern Himalayas and Nicobar Islands

1202. *Trichosteleum glauco-virens* (Mitt.) Broth.; **Present status:** Synonym of *Rhaphidostichum glauco-virens* (Mitt.) Broth.

Distribution in India: Eastern Himalayas (Endemic to India)

1203. *Trichosteleum hamatum* (Dozy & Molk.) A. Jaeger; **Present status:** Synonym of *Radulina hamata* (Dozy & Molk.) W.R. Buck & B.C. Tan

Distribution in India: Eastern Himalayas and South India

1204. *Trichosteleum luxurians* (Dozy & Molk.) Broth.; **Present status:** Synonym of *Papillidiopsis luxurians* (Dozy & Molk.) W.R. Buck & B.C. Tan

Distribution in India: Eastern Himalayas and Nicobar Islands

1205. *Trichosteleum monostictum* (Thwaites & Mitt.) Broth.; **Present status: Valid name**

Distribution in India: Eastern Himalayas, South India and Nicobar Islands

1206. *Trichosteleum punctipapillosum Paris ex Gangulee;* **Present status: Doubtful**

Distribution in India: Andaman Islands (Endemic to India)

1207. *Trichosteleum stereodontoides* Broth. ex Gangulee; **Present status: Doubtful**

Distribution in India: Eastern Himalayas

1208. *Trichosteleum stissophyllum* (Hampe & Müll. Hal.) A. Jaeger; **Present status:** Synonym of *Papillidiopsis stissophylla* (Hampe & Müll. Hal.) B.C. Tan & Y. Jia

Distribution in India: Eastern Himalayas

250. *Trolliella* Herzog

1209. *Trolliella euendostoma* Herzog; **Present status: Doubtful**

Distribution in India: Eastern Himalayas

250. *Warburgiella* C. Muell.

1210. *Warburgiella isopterygioides* Dixon & P. de la Varde; **Present status: Doubtful**

Distribution in India: South India

1211. *Warburgiella leptocarpa* (Schwägr.) M. Fleisch.; **Present status: Valid name**

Distribution in India: South India

1212. *Warburgiella leptorhynchoides* (Mitt.) M. Fleisch.; **Present status:** Synonym of *Rhaphidorrhynchium leptorhynchoides* (Mitt.) Broth.

Distribution in India: South India

1213. *Warburgiella perviridis* Dixon & P. de la Varde; **Present status: Doubtful**

Distribution in India: South India

251. *Wijkia* Crum

1214. *Wijkia baculifera* (Dixon) H.A. Crum ; **Present status: Valid name**

Distribution in India: Eastern Himalayas

1215. *Wijkia deflexifolia* (Mitt. ex Renauld & Cardot) H.A. Crum; **Present status: Valid name**

Distribution in India: Eastern Himalayas

1216. *Wijkia laxitexta* (Renauld & Cardot) H.A. Crum; **Present status: Valid name**

Distribution in India: Eastern Himalayas

1217. *Wijkia lepida* (Mitt.) H.A. Crum; **Present status: Valid name**

Distribution in India: Eastern Himalayas

1218. *Wijkia penicillata* (Mitt.) H.A. Crum; **Present status: Valid name**

Distribution in India: Eastern Himalayas

1219. *Wijkia sikkimense* (Broth.) R. S. Chopra; **Present status: Doubtful**

Distribution in India: Eastern Himalayas

1220. *Wijkia surcularis* (Mitt.) H.A. Crum; **Present status: Valid name**

Distribution in India: Eastern Himalayas

1221. *Wijkia tanytricha* (Mont.) H. A. Crum ; **Present status: Valid name**

Distribution in India: Western Himalayas, Eastern Himalayas and Central India

Family: Helodiaceae (M. Fleisch.) Ochyra

252. *Actinothuidium* (Besch.) Broth.

1222.. *Actinothuidium hookeri* (Mitt.) Broth.; **Present status: Valid name**

Distribution in India: Eastern Himalayas

Family: Symphyodontaceae M. Fleisch.

253. *Chaetomitriopsis* M. Fleisch.

1223. *Chaetomitriopsis glaucocarpa* (Reinw. ex Schwägr.) M. Fleisch.; **Present status: Valid name**

Distribution in India: Eastern Himalayas

254. *Chaetomitrium* Dozy & Molk.

1224. *Chaetomitrium papillifolium* Bosch & Sande Lac.; **Present status: Valid name**

Distribution in India: Eastern Himalayas

1225. *Chaetomitrium sikkimense* Broth. ex Gangulee; **Present status: Valid name**

Distribution in India: Eastern Himalayas

255. *Gammiella* Broth.

1226. *Gammiella pterogonioides* (Harv.) Paris; **Present status: Valid name**

Distribution in India: Eastern Himalayas

256. *Symphyodon* Mont.

1227. *Symphyodon asper* (Mitt.) A. Jaeger; **Present status: Valid name**

Distribution in India: Eastern Himalayas

1228. *Symphyodon complanatus* Dixon; **Present status: Valid name**

Distribution in India: Eastern Himalayas

1229. *Symphyodon echinatus* (Mitt.) A. Jaeger; **Present status: Valid name**

Distribution in India: Eastern Himalayas

1230. *Symphyodon erinaceus* (Mitt.) A. Jaeger; **Present status: Valid name**

Distribution in India: Eastern Himalayas

1231. *Symphyodon erraticus* (Mitt.) A. Jaeger; **Present status: Valid name**

Distribution in India: Eastern Himalayas

1232. *Symphyodon oblongifolius* (Renauld & Cardot) Broth.; **Present status: Valid name**

Distribution in India: Eastern Himalayas

1233. *Symphyodon orientalis* (Mitt.) Broth. ex Paris; **Present status: Valid name**

Distribution in India: Eastern Himalayas

1234. *Symphyodon perrottetii* Mont.; **Present status: Valid name**

Distribution in India: South India

1235. *Symphyodon scabrisetus* Hampe; **Present status:** Synonym of *Symphyodon erraticus* (Mitt.) A. Jaeger

Distribution in India: Eastern Himalayas

Family: Rhytidiaceae Broth.

257. *Rhytidium* (Sull.) Kindb.

1236. *Rhytidium rugosum* (Ehrh. & Hedw.) Kindb.; **Present status: Valid name**

Distribution in India: Western Himalayas

258. *Trigonodictyon* Dixon & P. de la Varde

1237. *Trigonodictyon indicum* Dixon & P. de la Varde; Present status: **Synonym of** *Dryptodon indicus* (Dixon & P. de la Varde) Ochyra & Żarnowiec

Distribution in India: South India

259. *Ulota* Mohr

1238. *Ulota robusta* Mitt.; **Present status: Doubtful**

Distribution in India: Eastern Himalayas

1239. *Ulota schmidii* (Müll. Hal.) Mitt.; **Present status: Valid name**

Distribution in India: South India

260. *Zygodon* Hook. & Taylor

1240. *Zygodon acutifolius* Müll. Hal.; **Present status:** Synonym of *Codonoblepharon acutifolium* (Müll. Hal.) A. Jaeger

Distribution in India: South India

1241. *Zygodon brevisetus* Wilson ex Mitt.; **Present status: Doubtful**

Distribution in India: Eastern Himalayas

1242. *Zygodon erosus* Mitt.; **Present status: Doubtful**

Distribution in India: South India

1243. *Zygodon humilis* Thwaites & Mitt.; **Present status:** Synonym of *Codonoblepharon pungens* (Müll. Hal.) A. Jaeger

Distribution in India: South India

1244. *Zygodon intermedius* Bruch & Schimp.; **Present status: Valid name**

Distribution in India: South India

1245. *Zygodon obtusifolius* Hook.; **Present status: Valid name**

Distribution in India: South India and Eastern Himalayas

1246. *Zygodon reinwardtii* (Hornsch.) A. Braun; **Present status: Valid name**

Distribution in India: South India

1247. *Zygodon tetragonostomus* A. Braun ex B.S.G.; **Present status: Valid name**

Distribution in India: South India

1248. *Zygodon viridissimus* (Dicks.) Brid.; **Present status: Valid name**

Distribution in India: Eastern Himalayas

1249. *Zygodon wightii* Dixon & Malta; **Present status: Doubtful**

Distribution in India: South India

N. ORDER: **POLYTRICHALES** M. FLEISCH

Family: Polytrichaceae Schwägr

261. ***Atrichum*** P. Beauv.

1250. *Atrichum aculeatum* Cardot & P. de la Varde ; Present status: **Synonym of** *Atrichum subserratum* (Harv. & Hook. f.) Mitt.

Distribution in India: Western Himalayas and South India

1251. *Atrichum flavisetum* Mitt.; **Present status: Valid name**

Distribution in India: Western Himalayas and Eastern Himalayas

1252. *Atrichum obtusulum* (Müll. Hal.) A. Jaeger ; **Present status: Valid name**

Distribution in India: Western Himalayas and Eastern Himalayas

1253. *Atrichum pallidum* Renauld & Cardot ; Present status: **Synonym of** *Atrichum subserratum* (Harv. & Hook. f.) Mitt.

Distribution in India: Western Himalayas and Eastern Himalayas

1254. *Atrichum undulatum* (Hedw.) P. Beauv. [Asthana & Sahu, 2013]; **Present status: Valid name**

Distribution in India: Western Himalayas and Eastern Himalayas

1255. *Atrichum undulatum* var. *subserratum* (Hook.) Paris ; Present **status:** Doubtful name

Distribution in India: Western Himalayas and Eastern Himalayas

262. *Lyellia* R. Br.

1256. *Lyellia bifurcata* Bél.; **Present status: Doubtful name**

Distribution in India: South India (Endemic to India)

1257. *Lyellia crispa* R. Br.; **Present status: Valid name**

Distribution in India: Eastern Himalayas

263. *Oligotrichum* Lam. & Cand.

1258. *Oligotrichum obtusatum* Broth.; Present **status:** Doubtful name

Distribution in India: Eastern Himalayas

1259. *Oligotrichum semilamellatum* (Hook. f.) Mitt.; **Present status: Valid name**

Distribution in India: Western Himalayas and Eastern Himalayas

264. *Pogonatum* P. Beauv.

1260. *Pogonatum akitense* Besch.; Present status: **Synonym of** *Pogonatum neesii* (Müll. Hal.) Dozy

Distribution in India: Eastern Himalayas

1261. *Pogonatum aloides* (Hedw.) P. Beauv.; **Present status: Valid name**

Distribution in India: Western Himalayas, Eastern Himalayas and South India

1262. *Pogonatum alpinum* (Hedw.) Röhl.; Present status: **Synonym of** *Polytrichastrum alpinum* (Hedw.) G.L. Sm.

Distribution in India: Western Himalayas and Eastern Himalayas

1263. *Pogonatum cirratum* (Sw.) Brid.; **Present status: Valid name**

Distribution in India: Eastern Himalayas

1264. *Pogonatum contortum* (Menzies ex Brid.) Lesq.; **Present status: Valid name**

Distribution in India: Eastern Himalayas

1265. *Pogonatum decolyi* Broth. ex Gangulee; **Present status: Synonym of** *Polytrichum patulum* Harv.

Distribution in India: Eastern Himalayas

1266. *Pogonatum fastigiatum* Mitt.; **Present status: Valid name**

Distribution in India: Eastern Himalayas

1267. *Pogonatum flexicaula* Mitt.; **Present status: Synonym of** *Pogonatum cirratum* (Sw.) Brid.

Distribution in India: Eastern Himalayas

1268. *Pogonatum fuscatum* Mitt.; **Present status: Valid name**

Distribution in India: Western Himalayas and Eastern Himalayas

1269. *Pogonatum grandifolium* (Lindb.) A. Jaeger; **Present status: Synonym of** *Pogonatum japonicum* Sull. & Lesq.

Distribution in India: Eastern Himalayas

1270. *Pogonatum gymnophyllum* Mitt. ; **Present status: Synonym of** Pogonatum proliferum (Griff.) Mitt.

Distribution in India: Eastern Himalayas

1271. *Pogonatum inflexum* (Lindb.) Sande Lac.; **Present status: Valid name**

Distribution in India: South India

1272. *Pogonatum hexagonum* Mitt.; Present status: **Synonym of** *Pogonatum patulum* (Harv.) Mitt.

Distribution in India: Western Himalayas, Eastern Himalayas and South India

1273. *Pogonatum himalayanum* Mitt.; Present status: **Synonym of** *Pogonatum urnigerum* (Hedw.) P. Beauv.

Distribution in India: Western Himalayas, Eastern Himalayas and South India

1274. *Pogonatum junghuhnianum* (Dozy & Molk.) Dozy & Molk.; Present status: **Synonym of** *Pogonatum neesii* (Müll. Hal.) Dozy

Distribution in India: Western Himalayas, Eastern Himalayas and South India

1275. *Pogonatum junghuhnianum* var. *sikkimense* Renauld & Cardot ; Present status: **Synonym of** *Pogonatum neesii* (Müll. Hal.) Dozy

Distribution in India: Western Himalayas, Eastern Himalayas and South India

1276. *Pogonatum leucopogon* Renauld & Cardot; Present status: **Synonym of** *Pogonatum neesii* (Müll. Hal.) Dozy

Distribution in India: Eastern Himalayas (Endemic to India)

1277. *Pogonatum macrophyllum* Dozy & Molk.; **Present status: Valid name**

Distribution in India: Eastern Himalayas

1278. *Pogonatum microstomum* (R. Br. & Schwägr.) Brid. [Asthana & Sahu, 2013]; **Present status: Valid name**

Distribution in India: Western Himalayas, Eastern Himalayas and South India

1279. *Pogonatum norkettii* R.S. Chopra; Present status: **Synonym of** *Pogonatum nudiusculum* Mitt.

Distribution in India: Eastern Himalayas

1280. *Pogonatum nudiusculum* Mitt.; Present status: **Present status: Valid name**

Distribution in India: Eastern Himalayas

1281. *Pogonatum papillosulum* Cardot & Dixon; Present status: **Synonym of** *Pogonatum neesii* (Müll. Hal.) Dozy

Distribution in India: Eastern Himalayas (Endemic to India)

1282. *Pogonatum perichaetiale* (Mont.) A. Jaeger [Asthana & Sahu, 2013]; Present status: **Synonym of** *Pogonatum gracilifolium* Besch.

Distribution in India: Western Himalayas, Eastern Himalayas and South India

1283. *Pogonatum proliferum* (Griff.) Mitt.; **Present status: Valid name**

Distribution in India: Eastern Himalayas

1284. *Pogonatum rufisetum* Mitt.; **Present status: Valid name**

Distribution in India: Eastern Himalayas

1285. *Pogonatum seminudum* Mitt.; Present status: **Synonym of** *Polytrichum proliferum* Griff.

Distribution in India: Western Himalayas and Eastern Himalayas

1286. *Pogonatum stevensii* Renauld & Cardot ; Present status: **Synonym of** *Pogonatum neesii* (Müll. Hal.) Dozy

Distribution in India: Western Himalayas and Eastern Himalayas

1287. *Pogonatum strictifolium* Broth. ex Gangulee; Present status: **Synonym of** *Pogonatum patulum* (Harv.) Mitt.

Distribution in India: Eastern Himalayas (Endemic to India)

1288. *Pogonatum subtortile* (Müll. Hal.) A. Jaeger; **Present status: Valid name**

Distribution in India: Eastern Himalayas and South India

1289. *Pogonatum subtortile* var. *teysmannianum* (Dozy & Molk.) Wijk & Margad.; **Present status: Valid name**

Distribution in India: Eastern Himalayas and South India

1290. *Pogonatum thomsonii* (Mitt.) A. Jaeger ; Present status: **Synonym of** *Pogonatum pericha&iale* subsp. *thomsonii* (Mitt.) Hyvönen

Distribution in India: Western Himalayas

1291. *Pogonatum tortipes* (Wilson ex Mitt.) A. Jaeger; **Present status: Valid name**

Distribution in India: Eastern Himalayas

1292. *Pogonatum urnigerum* (Hedw.) P. Beauv.; **Present status: Valid name**

265. *Polytrichastrum* G. L. Sm.

1293. *Polytrichastrum papillatum* G. L. Sm.; **Present status: Valid name**

Distribution in India: Western Himalayas

1294. *Polytrichastrum xanthopilum* (Wilson ex Mitt.) G.L. Sm.; **Present status: Valid name**

Distribution in India: Eastern Himalayas

266. *Polytrichum* Hedw.

1295. *Polytrichum alpinum* Hedw.; Present status: **Synonym of** *Polytrichastrum alpinum* (Hedw.) G.L. Sm.

Distribution in India: Western Himalayas and Eastern Himalayas

1296. *Polytrichum densifolium* Wilson *ex* Mitt.; Present status: **Synonym of** *Polytrichastrum formosum* var. *densifolium* (Wilson ex Mitt.) Z. Iwats. & Nog.

Distribution in India: Eastern Himalayas

1297. *Polytrichum juniperinum* Hedw.; **Present status: Valid name**

Distribution in India: Western Himalayas

1298. *Polytrichum piliferum* Hedw.; Present status: **Synonym of** *Polytrichum juniperinum* Hedw.

Distribution in India: Western Himalayas

O. ORDER: POTTIALES M. FLEISCH.

Family: Calymperaceae Kindb.

267. *Calymperes* Sw.

1299. *Calymperes afzelii* Sw.; **Present status: Valid name**

Distribution in India: South India

1300. *Calymperes andamense* Besch.; Present status: **Synonym of** *Calymperes tahitense* (Sull.) Mitt.

Distribution in India: Andaman Islands (Endemic to India)

1301. *Calymperes boulayi* Besch.; **Present status: Valid name**

Distribution in India: South India

1302. *Calymperes delessertii* Besch.; Present status: **Synonym of** *Calymperes boulayi* Besch.

Distribution in India: Andaman Islands (Endemic to India)

1303. *Calymperes hampei* Dozy & Molk; Present status: **Synonym of** *Calymperes erosum* Müll. Hal.

Distribution in India: Eastern Himalayas

1304. *Calymperes kurzianum* Hampe ex Müll. Hal.; Present status: **Synonym of** *Calymperes moluccense* Schwägr.

Distribution in India: Andaman Islands

1305. *Calymperes linguatum* Müll. Hal. ex Besch.; Present status: **Doubtful name**

Distribution in India: Andaman Islands (Endemic to India)

1306. *Calymperes lonchophyllum* Schwagr. ; Present status: **Doubtful name**

Distribution in India: South India

1307. *Calymperes mangalorense* Dixon & P. de la Varde; **Present status:** Synonym of *Calymperes erosum* Müll. Hal.

Distribution in India: South India

1308. *Calymperes manii* Müll. Hal. ex Besch.; Present status: **Synonym of** *Calymperes erosum* Müll. Hal.

Distribution in India: South India, Andaman Islands

1309. *Calymperes motley* Mitt.; **Present status: Valid name**

Distribution in India: South India

1310. *Calymperes nietneri* var. *atro-viride* Dixon; **Present status: Synonym of** Synonym of *Calymperes lonchophyllum* Schwägr.

Distribution in India: South India

1311. *Calymperes nicobarense* Hampe; **Present status: Synonym of** *Calymperes graeffeanum* Müll. Hal.

Distribution in India: Nicobar Islands

1312. *Calymperes nukahivense* Besch.; Present status: **Synonym of** *Calymperes graeffeanum* Müll. Hal.

Distribution in India: Indian Orientalis

1313. *Calymperes omanicum* Besch.; **Present status: Synonym of** Synonym of *Calymperes motleyi* Mitt.

Distribution in India: South India

1314. *Calymperes pachyphyllum* Thér. & P. de la Varde; **Present status**: **Synonym of** is a Synonym of *Calymperes beccarii* Hampe

Distribution in India: South India

1315. *Calymperes palisotii* Schwägr.; **Present status**: **Valid name**

Distribution in India: South India

1316. *Calymperes punctulatum* Hampe; Present status: **Synonym of** *Calymperes graeffeanum* Müll. Hal.

Distribution in India: Nicobar Islands

1317. *Calymperes sikkimense* Gangulee; Present status: **Synonym of** *Heliconema peguense* (Besch.) L.T. Ellis & A. Eddy

Distribution in India: Eastern Himalayas (Endemic to India)

1318. *Calymperes sundarbanense* Gangulee; Present status: Synonym of *Heliconema peguense* (Besch.) L.T. Ellis & A. Eddy

Distribution in India: Gangetic plains (Endemic to India)

1319. *Calymperes tenerum* Mull. Hull.; **Present status: Valid name**

Distribution in India: Gangetic plains and South India

1320. *Calymperes thwaitesii* subs. *fordii* Besch.; **Present status:** Synonym of *Calymperes afzelii* Sw.

Distribution in India: South India

1321. *Calymperes tortelloides* Broth. & Dixon; **Present status:** Synonym of *Trichostomum tortelloides* (Broth. & Dixon) R.H. Zander

Distribution in India: South India

1322. *Calymperes vriesei* Besch.; **Present status: Valid name**

Distribution in India: Eastern Himalayas

268. ***Exodictyon*** Cardot

1323. *Exodictyon blumii* (Nees ex Hampe) M. Fleisch.; **Present status: Synonym of** *Exostratum blumii* (Nees ex Hampe) L.T. Ellis

Distribution in India: Eastern Himalayas and Nicobar Island

269. ***Heliconema*** (Mitt.) L.T. Ellis & A. Eddy

1324. *Heliconema peguense* (Besch.) L.T. Ellis & A. Eddy; **Present status: Valid name**

Distribution in India: South India

270. ***Leucophanes*** Brid.

1325. *Leucophanes albescens* Müll. Hal.; **Present status: Synonym of** *Leucophanes glaucum* (Schwägr.) Mitt.

Distribution in India: Nicobar Island

1326. *Leucophanes glaucescens* Müll. Hal. ex M. Fleisch.; **Present status: Synonym of** *Leucophanes glaucum* (Schwägr.) Mitt.

Distribution in India: Andaman Island

1327. *Leucophanes nicobaricum* Müll. Hal.; **Present status: Synonym of** *Leucophanes octoblepharioides* Brid.

Distribution in India: Nicobar Island (Endemic to India)

1328. *Leucophanes octoblepharioides* Brid.; **Present status: Valid name**

Distribution in India: Eastern Himalayas

Family: Ephemeraceae Schimp.

271. *Ephemerum* Hampe

1329. *Ephemerum asiaticum* Paris & Broth.; **Present status: Doubtful**

Distribution in India: South India

272. *Micromitrium* Austin

1330. *Micromitrium tenerum* (Bruch & Schimp.) Crosby; **Present status: Valid name**

Distribution in India: Eastern Himalayas and South India

273. *Mitthyridium* H. Rob.

1331. *Mitthyridium cardotii* (M. Fleisch.) H. Rob; **Present status: Synonym of** *Mitthyridium fasciculatum* subsp. *cardotii* (M. Fleisch.) B.C. Tan & L.T. Ellis

Distribution in India: Eastern Himalayas

1332. *Mitthyridium fasciculatum* (Hook. & Grev.) H. Rob.; **Present status: Valid name**

Distribution in India: South India

1333. *Mitthyridium flavum* (Müll. Hal.) H. Rob.; **Present status: Valid name**

Distribution in India: Andaman and Nicobar Islands

1334. *Mitthyridium obtusifolium* (Lindb.) H. Rob.; **Present status: Valid name**

Distribution in India: Andaman Islands

1335. *Mitthyridium repens* (Harv.) H. Rob.; **Present status: Valid name**

Distribution in India: South India and Andaman Islands

1336. *Mitthyridium undulatum* (Dozy & Molk.) H. Rob.; **Present status: Valid name**

Distribution in India: Andaman Islands

Family: Pottiaceae Schimp.

274. *Aloina* (Müll. Hal.) Kindb.

1337. *Aloina rigida* (Hedw.) Limpr.; **Present status: Valid name**

Distribution in India: Western Himalayas

275. *Anoectangium* Schwägr.

1338. *Anoectangium euchloron* (Schwägr.) Mitt.; Present status: **Synonym of** *Anoectangium aestivum* (Hedw.) Mitt.

Distribution in India: South India

1339. *Anoectangium bicolor* Renauld & Cardot ; Present status: **Doubtful name**

Distribution in India: Western Himalayas and Eastern Himalayas

1340. *Anoectangium clarum* Mitt.; **Present status: Valid name**

Distribution in India: Western Himalayas and Eastern Himalayas

1341. *Anoectangium kashmiriense* M. N. Aziz & Vohra ; Present status: **Doubtful name**

Distribution in India: Western Himalayas

1342. *Anoectangium sikkimense* M.N. Aziz & Vohra; Present status: **Doubtful name**

Distribution in India: Eastern Himalayas

1343. *Anoectangium stracheyanum* Mitt.; **Present status: Valid name**

Distribution in India: Western Himalayas, Eastern Himalayas and South India

1344. *Anoectangium thomsonii* Mitt.; **Present status: Valid name**

Distribution in India: Western Himalayas, Eastern Himalayas and South India

1345. *Anoectangium walkeri* Broth.; Present status: **Doubtful name**

Distribution in India: Western Himalayas and South India

276. *Astomum* Hampe

1346. *Astomum crispum* (Hedw.) Hampe ; **Present status: Valid name**

Distribution in India: Western Himalayas

1346. *Astomum minutum* Dixon & P. de la Varde; **Present status:** Synonym of *Weissia minuta* (Dixon & P. de la Varde) M.N. Aziz & Vohra

Distribution in India: South India

277. *Barbula* Hedw.

1347. *Barbula arcuata* Griff.; **Present status: Valid name**

Distribution in India: Western Himalayas, Eastern Himalayas, Central India, Gangetic plains and South India

1348. *Barbula confertifolia* Mitt.; *Present status*: **Synonym of** *Didymodon confertifolius* (Mitt.) Paris

Distribution in India: Western Himalayas

1349. *Barbula dharvarensis* Dixon; Present status: **Doubtful name**

Distribution in India: South India

1350. *Barbula funalis* Dixon & Badhw.; Present status: **Doubtful name**

Distribution in India: Western Himalayas

1351. *Barbula fuscescens* Wall.; **Present status: Valid name**

Distribution in India: Western Himalayas

1352. *Barbula gracilenta* Mitt.; **Present status: Valid name**

Distribution in India: Western Himalayas, Eastern Himalayas and Gangetic plains

1353. *Barbula gregaria* (Mitt.) A. Jaeger ; *Present status*: **Synonym of** *Barbula indica* var. *gregaria* (Mitt.) R.H. Zander

Distribution in India: Western Himalayas and Eastern Himalayas

1354. *Barbula horricomis* Müll. Hal. & Gangulee ; Present status: **Doubtful name**

Distribution in India: Western Himalayas and Eastern Himalayas

1355. *Barbula indica* (Hook.) Spreng.; **Present status: Valid name**

Distribution in India: Western Himalayas, Eastern Himalayas, Central India, Gangetic plains, Rajasthan and South India

1356. *Barbula leucodontoides* (Gangulee) Gangulee ; **Present status: Valid name**

Distribution in India: Western Himalayas and Eastern Himalayas

1357. *Barbula marginatula* Gangulee; Present status: **Doubtful name**

Distribution in India: Eastern Himalayas

1358. *Barbula microstoma* (Dixon & Badhw.) R. S. Chopra ; **Present status: Valid name**

Distribution in India: Western Himalayas

1359. *Barbula obscura* Mitt.; Present status**: Synonym of** *Didymodon mittenii* Gangulee

Distribution in India: Western Himalayas

1360. *Barbula ovata* Mitt.; Present status**: Synonym of** *Trichostomum ovatum* (Mitt.) Müll. Hal.

Distribution in India: Western Himalayas

1361. *Barbula reflexa* (Brid.) Brid.; Present status: **Synonym of** *Didymodon ferrugineus* (Schimp. & Besch.) M.O. Hill

Distribution in India: Western Himalayas and Eastern Himalayas

1362. *Barbula spathulifolia* **(Dixon & P. de la Varde) R.H. Zander; Present status: Valid name**

Distribution in India: India Orientalis

1363. *Barbula subcontorta* Broth.; *Present status***: Synonym of** *Didymodon vinealis* (Brid.) R.H. Zander

Distribution in India: Western Himalayas

1364. *Barbula tenuirostris* Brid.; **Present status: Valid name**

Distribution in India: Eastern Himalayas and Gangetic plains

1365. *Barbula unguiculata* Hedw.; **Present status: Valid name**

Distribution in India: Western Himalayas and Eastern Himalayas

1366. *Barbula vardei* R. S. Chopra; **Present status: Valid name**

Distribution in India: India Orientalis

1367. *Barbula vinealis* Brid.; *Present status*: **Synonym of** *Didymodon vinealis* (Brid.) R.H. Zander

278. *Bellibarbula* P. C. Chen

1368. *Bellibarbula kurziana* Hampe ex P.C. Chen; **Present status: Doubtful name**

Distribution in India: Eastern Himalayas

1369. *Bellibarbula kruziana* var. *purpurascens* Gangulee; **Present status: Doubtful name**

Distribution in India: Eastern Himalayas

279. *Bryoerythrophyllum* P. C. Chen

1370. *Bryoerythrophyllum alpigenum* (Vent.) P.C. Chen ; **Present status: Valid name**

Distribution in India: Western Himalayas

1371. *Bryoerythrophyllum atrorubens* (Besch.) P.C. Chen ; **Present status: Valid name**

Distribution in India: Western Himalayas

1372. *Bryoerythrophyllum ferrugineum* Gangulee; Present status: **Synonym of** *Bryoerythrophyllum ferruginascens* (Stirt.) Giacom.

Distribution in India: Eastern Himalayas (Endemic to India)

1373. *Bryoerythrophyllum gymnostomum* (Broth.) P. C. Chen ; **Present status: Valid name**

Distribution in India: Western and Eastern Himalayas

1374. *Bryoerythrophyllum inaequalifolium* (Taylor) R. H. Zander; **Present status: Valid name**

Distribution in India: Western Himalayas and South India

1375. *Bryoerythrophyllum recurvirostrum* (Hedw.) P. C. Chen; **Present status: Valid name**

Distribution in India: Western Himalayas and South India

1376. *Bryoerythrophyllum yunnanense* (Herzog) P.C. Chen; Present status: **Synonym of** *Erythrobarbula yunnanensis* (Herzog) Steere

Distribution in India: Western and Eastern Himalayas

1377. *Bryoerythrophyllum yunnanense* var. *noguchianum* Gangulee; Present status: **Synonym of** *Bryoerythrophyllum noguchianum* (Gangulee) K. Saito

Distribution in India: Western and Eastern Himalayas

280. *Cinclidotus* P. Beauv.

1378. *Cinclidotus acutifolius* Broth.; Present status: **Doubtful name**

Distribution in India: Western Himalayas

281. *Desmatodon* Brid.

1379. *Desmatodon cernuus* (Huebener) Bruch & Schimp.; **Present status: Valid name**

Distribution in India: Western Himalayas

1380. *Desmatodon gemmascens* P. C. Chen ; **Present status: Valid name**

Distribution in India: Western Himalayas and Eastern Himalayas

1381. *Desmatodon kabir-khanii* Broth.; Present status: **Synonym of** *Tortula kabir-khanii* (Broth.) R.H. Zander

Distribution in India: Western Himalayas

1382. *Desmatodon leucostoma* (R. Br.) Berggr.; Present status: **Synonym of** *Tortula leucostoma* (R. Br.) Hook. & Grev.

Distribution in India: Western Himalayas

282. *Didymodon* Hedw.

1383. *Didymodon asperifolius* (Mitt.) H.A. Crum, Steere & L.E. Anderson ; **Present status: Valid name**

Distribution in India: Western Himalayas and Eastern Himalayas

1384. *Didymodon canaliculatus* Dixon ; Present status: **Synonym of** *Barbula canaliculata* (Dixon) R.S. Chopra

Distribution in India: Western Himalayas

1385. *Didymodon constrictus* (Mitt.) K. Saito ; **Present status: Valid name**

Distribution in India: Western Himalayas and Eastern Himalayas

1386. *Didymodon dixonii* Wadhwa & Vohra; Present status: **Synonym of** *Barbula divergens* Broth.

Distribution in India: South India

1387. *Didymodon fallax* (Hedw.) R. H. Zander ; **Present status: Valid name**

Distribution in India: Western Himalayas

1388. *Didymodon hastatus* (Mitt.) R. H. Zander ; **Present status: Valid name**

Distribution in India: Western Himalayas and Eastern Himalayas

1389. *Didymodon leskeoides* K. Saito ; Present status: **Doubtful name**

Distribution in India: Western Himalayas

1390. *Didymodon maschalogena* (Renauld & Cardot) Broth.; **Present status: Valid name**

Distribution in India: Eastern Himalayas (Endemic to India)

1391. *Didymodon michiganensis* (Steere) K. Saito; **Present status: Synonym of** *Didymodon maschalogena* (Renauld & Cardot) Broth.

Distribution in India: Eastern Himalayas

1392. *Didymodon mittenii* Gangulee ; **Present status: Valid name**

Distribution in India: Western Himalayas and Eastern Himalayas

1393. *Didymodon nigrescens* (Mitt.) K. Saito ; **Present status: Valid name**

Distribution in India: Western Himalayas and Eastern Himalayas

1394. *Didymodon recurvus* (Griff.) Broth.; **Present status: Valid name**

Distribution in India: South India

1395. *Didymodon rigidulus* Hedw.; **Present status: Valid name**

Distribution in India: Western Himalayas and Eastern Himalayas and South India

1396. *Didymodon rigidulus* var. *gracilis* (Schleich. & Hook. & Grev.) R.H. Zander ; **Present status: Valid name**

Distribution in India: Western Himalayas, Eastern Himalayas and South India

1397. *Didymodon stewartii* (E.B. Bartram) R.H. Zander ; **Present status: Valid name**

Distribution in India: Western Himalayas

1398. *Didymodon strictifolius* Dixon & P. de la Varde; **Present status:** Synonym of *Bellibarbula recurva* (Griff.) R.H. Zander

Distribution in India: South India

1399. *Didymodon tophaceus* (Brid.) Lisa ; **Present status: Valid name**

Distribution in India: Western Himalayas

1400. *Didymodon vinealis* (Brid.) R. H. Zander ; **Present status: Valid name**

Distribution in India: Western Himalayas and Eastern Himalayas

283. *Eucladium* Bruch & Schimp.

1401. *Eucladium verticillatum* (Hedw.) Bruch & Schimp.; **Present status: Valid name**

Distribution in India: Western Himalayas

284. *Ganguleea* R. H. Zander

1402. *Ganguleea angulosa* (Broth. & Dixon) R. H. Zander ; **Present status: Valid name**

Distribution in India: Western Himalayas and Eastern Himalayas

285. *Gymnostomum* Nees & Hornsch.

1403. *Gymnostomum aeruginosum* G. L. Sm.; **Present status: Valid name**

Distribution in India: Western Himalayas

1404. *Gymnostomum calcareum* Nees & Hornsch.; Present status: **Synonym of** *Gymnostomum aeruginosum* Sm.

Distribution in India: Western Himalayas and Eastern Himalayas

286. *Hydrogonium* (Müll. Hal.) A. Jaeger.

1405. *Hydrogonium amplexifolium* (Mitt.) P. C. Chen ; **Present status: Valid name**

Distribution in India: Western Himalayas and Eastern Himalayas

1406. *Hydrogonium arcuatum* (Griff.) Wijk & Margad.; **Present status: Valid name**

Distribution in India: Western Himalayas, Eastern Himalayas, South India, Central India, Gangetic plains and Rajasthan

1407. *Hydrogonium consanguineum* (Thwaites & Mitt.) Hilp.; **Present status: Valid name**

Distribution in India: Western Himalayas, Eastern Himalayas, South India, Central India, Gangetic plains and Rajasthan

1408. *Hydrogonium decolyi* Gangulee; Present status: Synonym of *Bellibarbula recurva* (Griff.) R.H. Zander

Distribution in India: Western Himalayas and Eastern Himalayas

1409. *Hydrogonium dicranelloides* Gangulee ; Present status: **Doubtful name**

Distribution in India: Western Himalayas and Eastern Himalayas

1410. *Hydrogonium ehrenbergii* (Lorentz) A. Jaeger.; **Present status: Valid name**

Distribution in India: Western Himalayas

1411. *Hydrogonium gracilentum* (Mitt.) P. C. Chen ; **Present status: Valid name**

Distribution in India: Western Himalayas, Eastern Himalayas and Central India

1412. *Hydrogonium javanicum* (Dozy & Molk.) Hilp.; **Present status: Valid name**

Distribution in India: Western Himalayas, Eastern Himalayas, Panjab, Rajasthan and Gangetic plains

1413. *Hydrogonium mussoorianum* Vohra ; Present status: **Doubtful name**

Distribution in India: Western Himalayas

1414. *Hydrogonium pseudo-ehrenbergii* (Fleisch.) P. C. Chen; Present status: **Doubtful name**

Distribution in India: Western Himalayas and South India

1415. *Hydrogonium subpellucidum* (Mitt.) Hilp.; **Present status: Valid name**

Distribution in India: Western Himalayas and Eastern Himalayas

287. *Hymenostomum* R. Br.

1416. *Hymenostomum edentulum* (Mitt.) Besch.; **Present status: Valid name**

Distribution in India: South India and Central India

1417. *Hymenostomum microstomum* (Hedw.) R. Br.; Present status**: Synonym of** *Weissia brachycarpa* (Nees & Hornsch.) Jur.

Distribution in India: Western Himalayas

1418. *Hymenostomum rostellatum* (Brid.) Schimp.; Present status**: Synonym of** *Weissia rostellata* (Brid.) Lindb.

Distribution in India: Western Himalayas

1419. *Hymenostomum tortile* (Schwägr.) Bruch & Schimp.; Present status**: Synonym of** *Weissia condensa* (Voit) Lindb.

Distribution in India: Western Himalayas

288. *Hymenostyliella* E. B. Bartram

1420. *H. llanosii* (Broth.) H. Rob.; **Present status: Valid name**

Distribution in India: Western Himalayas and Central India

289. *Hymenostylium* Brid.

1421. *Hymenostylium annotinum* Mitt. ex Dixon; Present status: **Doubtful name**

Distribution in India: India Orientalis

1422. *Hymenostylium dicranelloides* Broth. & Dixon ; **Present status: Valid name**

Distribution in India: Western Himalayas

1423. *Hymenostylium filiforme* Dixon ; Present status**: Synonym of** *Didymodon filiforme* (Dixon) M.N. Aziz & Vohra

Distribution in India: Western Himalayas

1424. *Hymenostylium grandirete* Dixon ; Present status: **Doubtful name**

Distribution in India: Western Himalayas

1425. *Hymenostylium recurvirostrum* (Hedw.) Dixon [Asthana & Sahu, 2013]; **Present status: Valid name**

Distribution in India: Western Himalayas, Eastern Himalayas, South India, Punjab

1426. *Hymenostylium recurvirostrum* var. *aurantiacum* (Mitt.) Gangulee; **Present status: Valid name**

Distribution in India: Western Himalayas, Eastern Himalayas and South India

1427. *Hymenostylium shepheardae* Cardot & Dixon ; Present status: **Synonym of** *Anoectangium shepherdae* (Cardot & Dixon) R.H. Zander

Distribution in India: Western Himalayas

1428. *Hymenostylium validinerve* Dixon et P. Vard.; Present status: **Doubtful name**

Distribution in India: South India

290. *Hyophila* Brid.

1429. *Hyophila comosa* Dixon; Present status: **Synonym of** *Hyophila rosea* R.S. Williams

Distribution in India: South India, Rajasthan and Central India

1430. *Hyophila involuta* (Hook.) A. Jaeger.; **Present status: Valid name**

Distribution in India: Western Himalayas, Eastern Himalayas, Rajasthan, Central India, South India, Gangetic Plains

1431. *Hyophila kurziana* Gangulee; **Present status: Doubtful name**

Distribution in India: Eastern Himalayas (Endemic to India)

1432. *Hyophila mollifolia* Dixon & P. de la Varde; **Present status: Doubtful name**

Distribution in India: South India

1433. *Hyophila perannulata* Renauld & Cardot; Present status: **Synonym of** *Trichostomum criotum* R.H. Zander

Distribution in India: Eastern Himalayas (Endemic to India)

1434. *Hyophila rosea* R. S. Williams ; **Present status: Valid name**

Distribution in India: Western Himalayas, Rajasthan, Central India and South India,

1435. *Hyophila spathulata* (Harv.) A. Jaeger.; Present status**: Synonym of** *Tortula spathulata* (Harv.) Mitt.

Distribution in India: Western Himalayas and Eastern Himalayas

291. *Hyophilopsis* Cardot *et* Dixon

1436. *Hyophilopsis entosthodontacea* Cardot & Dixon; Present status**: Synonym of** *Tortula entosthodontacea* (Cardot & Dixon) R.H. Zander

Distribution in India: South India

292. *Leptodontium* (Müll. Hal.) Hampe

1437. *Leptodontium flexifolium* (Dicks.) Hampe ; **Present status: Valid name**

Distribution in India: Western Himalayas and Eastern Himalyas

1438. *Leptodontium handelii* Thér; **Present status: Doubtful name**

Distribution in India: Eastern Himalyas

1439. *Leptodontium viticulosoides* (P. Beauv.) Wijk & Margad.; **Present status: Valid name**

Distribution in India: Eastern Himalyas

293. *Molendoa* Lindb.

1440. *Molendoa duthiei* (Broth.) Broth.; **Present status: Valid name**

Distribution in India: Western Himalayas

1441. *Molendoa roylei* (Mitt.) Broth.; **Present status: Valid name**

Distribution in India: Western Himalayas

1442. *Molendoa sendtneriana* (Bruch & Schimp.) Limpr.; **Present status: Valid name**

Distribution in India: Western Himalayas

294. ***Oxystegus*** (Limpr.) Hilp.

1443. *Oxystegus burmensis* (E.B. Bartram) Gangulee; **Present status: Valid name**

Distribution in India: Indian Orientalis

1444. *Oxystegus cylindricus* (Bruch & Brid.) Hilp.; Present status**: Synonym of** *Trichostomum tenuirostre* var. *tenuirostre*

Distribution in India: Western Himalayas, Eastern Himalayas and South India

1445. *Oxystegus cylindrothecus* (Mitt.) Gangulee; **Present status: Valid name**

Distribution in India: Eastern Himalayas (Endemic to India)

1446. *Oxystegus indicus* (Dixon & P. de la Varde) Hilp.; **Present status: Valid name**

Distribution in India: South India

1447. *Oxystegus khasianus* (Mitt.) Gangulee; Present status**: Synonym of** *Pseudosymblepharis khasiana* (Mitt.) R.H. Zander

Distribution in India: Eastern Himalayas (Endemic to India)

1448. *Oxystegus stenophyllus* (Mitt.) Gangulee ; **Present status: Valid name**

Distribution in India: Western Himalayas, Eastern Himalayas and South India

1449. *Oxystegus tenuirostris* (Hook. & Taylor) A.J.E. Sm. [Asthana & Sahu, 2013]; Present status**: Synonym of** *Trichostomum tenuirostre* (Hook. & Taylor) Lindb.

Distribution in India: Western Himalayas, Eastern Himalayas and South India

1450. *Oxystegus tenuirostris* var. *gemmiparus* (Schimp.) R. H. Zander ; Present status: **Doubtful name**

Distribution in India: Western Himalayas

295. ***Pleurochaete*** Lindb.

1451. *Pleurochaete squarrosa* (Brid.) Lindb.; Present **status: Valid name**

Distribution in India: Western Himalayas and South India

296. ***Pottia*** (Reichenb.) Fuernr.

1452. *Pottia alpicola* Dixon ; Present status: **Doubtful name**

Distribution in India: Western Himalayas

1453. *Pottia starckeana* (Hedw.) Müll. Hal.; Present status**: Synonym of** *Microbryum starckeanum* (Hedw.) R.H. Zander

Distribution in India: Western Himalayas

1454. *Pottia watsonii* R. S. Chopra; Present status: **Doubtful name**

Distribution in India: South India

297. *Pseudocrossidium* Williams

1455. *Pseudocrossidium porphyreoneurum* (Müll. Hal.) R.H. Zander; **Present status: Valid name**

Distribution in India: Rajasthan

298. *Reimersia* P. C. Chen

1456. *Reimersia inconspicua* (Griff.) P. C. Chen ; **Present status: Valid name**

Distribution in India: Western Himalayas and Eastern Himalayas

299. *Rhamphidium* Mitt.

1457. *Rhamphidium cribbifolium* Dixon & Wijk, Margad. & Florsch.; Present status: **Doubtful name**

Distribution in India: Western Himalayas

1458. *Rhamphidium laticuspe* Dixon ; Present status: **Doubtful name**

Distribution in India: Western Himalayas

1459. *Rhamphidium madurense* Dixon & P. de la Varde; Present status: **Doubtful name**

Distribution in India: South India

1460. *Rhamphidium mussuriensis* Dixon; Present status: **Doubtful name**

Distribution in India: Western Himalayas

300. *Scopelophila* (Mitt.) Lindb.

1461. *Scopelophila cataractae* (Mitt.) Broth.; **Present status: Valid name**

Distribution in India: Western Himalayas

1462. *Scopelophila ligulata* (Spruce) Spruce ; **Present status: Valid name**

Distribution in India: Western Himalayas

301. *Semibarbula* Herz. & Hilp.

1463. *Semibarbula orientalis* (F. Weber) Wijk & Margad.; **Present status: Valid name**

Distribution in India: Western Himalayas, Eastern Himalayas, Central India and Gangetic plains

1464. *Semibarbula ranuii* Gangulee; **Present status: Doubtful**

Distribution in India: Central India and Gangetic plains

302. *Stegonia* Vent.

1465. *Stegonia latifolia* (Schwägr.) Venturi & Broth.; **Present status: Valid name**

Distribution in India: Western Himalayas

303. *Syntrichia* Brid.

1466. *Syntrichia princeps* (De Not.) Mitt.; **Present status: Valid name**

Distribution in India: Western Himalayas and Eastern Himalayas

304. *Timmiella* (De Not.) Limpr.

1467. *Timmiella anomala* (Bruch & Schimp.) Limpr.; **Present status: Valid name**

Distribution in India: Western Himalayas and South India

1468. *Timmiella barbuloides* (Brid.) Mönk.; **Present status: Valid name**

Distribution in India: Western Himalayas

1469. *Timmiella diminuta* (Müll. Hal.) P. C. Chen ; **Present status: Valid name**

Distribution in India: Western Himalayas

305. *Tortella* (Lindb.) Limpr.

1476. *Tortella alpicola* Dixon; **Present status: Valid name**

Distribution in India: Western Himalayas

1470. *Tortella ceylonensis* M. Fleisch. ex Dixon; **Present status:** Synonym of *Tortula humilis* (Hedw.) Turner

Distribution in India: Indian Orientalis

1471. *Tortella fragilis* (Hook. & Wilson) Limpr.; **Present status: Valid name**

Distribution in India: Western Himalayas and Eastern Himalayas

1472. *Tortella tortuosa* (Hedw.) Limpr.; **Present status: Valid name**

Distribution in India: Western Himalayas

1473. *Tortella walkeri* (Broth.) R.H. Zander; **Present status: Valid name**

Distribution in India: Indian Orientalis

306. *Tortula* Hedw.

1474. *Tortula brandisii* (Müll. Hal.) Broth.; Present status**: Synonym of** *Syntrichia brandisii* (Müll. Hal.) R.H. Zander

Distribution in India: Western Himalayas

1475. *Tortula caninervis* (Mitt.) Broth. ; Present status: Synonym of Synonym of *Syntrichia caninervis* Mitt.

Distribution in India: Eastern Himalayas

1476. *Tortula inermis* (Brid.) Mont.; **Present status: Valid name**

Distribution in India: Western Himalayas

1477. *Tortula microphylligera* Broth.; **Present status: Doubtful**

Distribution in India: Eastern Himalayas

1478. *Tortula montana* Mitt.; Present status**: Synonym of** *Rhamphidium montanus* (Mitt.) R.H. Zander

Distribution in India: Western Himalayas

1479. *Tortula mucronifolia* Schwaegr.; **Present status: Valid name**

Distribution in India: India Orientalis

1480. *Tortula muralis* Hedw.; **Present status: Valid name**

Distribution in India: Western Himalayas and Eastern Himalayas

1481. *Tortula nigra* R.H. Zander; Present status**: Synonym of** *Syntrichia percarnosa* (Müll. Hal.) R.H. Zander

Distribution in India: India Orientalis

1482. *Tortula princeps* De Not.; Present status**: Synonym of** *Syntrichia princeps* (De Not.) Mitt.

Distribution in India: Western Himalayas and Eastern Himalayas

1483. *Tortula. pseudo-princeps* Dixon ; Present status: **Doubtful name**

Distribution in India: Western Himalayas

1484. *Tortula rubripila* Dixon ; Present status**: Synonym of** *Tortula norvegica* var. *norvegica*

Distribution in India: Western Himalayas

1485. *Tortula ruralis* (Hedw.) P. Gaertn., B. Mey. & Scherb.; *Present status*: **Synonym of** *Syntrichia ruralis* (Hedw.) F. Weber & D. Mohr

Distribution in India: Western Himalayas

1486. *Tortula schmidii* (Müll. Hal.) Broth.; **Present status: Valid name**

Distribution in India: South India

1487. *Tortula subulata* Hedw.; **Present status: Valid name**

Distribution in India: Western Himalayas and Eastern Himalayas

1488. *Tortula websteri* H. Rob.; Present status: **Doubtful name**

Distribution in India: Western Himalayas and Eastern Himalayas

307. *Trichostomum* Bruch

1489. *Trichostomum bombayense* Müll. Hal.; **Present status: Valid name**

Distribution in India: Central India

1490. *Trichostomum brachydontium* Müll. Hal.; **Present status: Valid name**

Distribution in India: Western Himalayas and Eastern Himalayas

1491. *Trichostomum hyalinoblastum* (Broth.) Broth.; **Present status: Valid name**

Distribution in India: South India

1492. *Trichostomum lillei* Dixon ; Present status: **Doubtful name**

Distribution in India: Western Himalayas

1493. *Trichostomum minusculum* Dixon & P. de la Varde; Present status: Synonym of *Pseudosymblepharis bombayensis* (Müll. Hal.) P. Sollman

Distribution in India: South India

1494. *Trichostomum orthodontum* (Müll. Hal.) Broth.; **Present status: Valid name**

Distribution in India: South India

1495. *Trichostomum perannulatum* Dixon & P. de la Varde; **Present status:** Synonym of *Pseudosymblepharis bombayensis* (Müll. Hal.) P. Sollman

Distribution in India: South India

1496. *Trichostomum subminusculum* Dixon & P. de la Varde; **Present status:** Synonym of *Pseudosymblepharis bombayensis* (Müll. Hal.) P. Sollman

Distribution in India: South India

1497. *Trichostomum uncinifolium* Dixon; Present status: **Synonym of** *Pseudosymblepharis bombayensis* (Müll. Hal.) P. Sollman

Distribution in India: Western Himalayas

308. *Trachycarpidium* Broth.

1498. *Trachycarpidium tisserantii* Dixon & P. de la Varde; Present status: **Doubtful name**

Distribution in India: South India

309. *Trachyphyllum* Gepp *in* Hiern.

1499. *Trachyphyllum elongatum* Dixon & P. de la Varde; Present status: Synonym of *Schwetschkeopsis elongata* (Dixon & P. de la Varde) W.R. Buck & H.A. Crum

Distribution in India: India Orientalis

1500. *Trachyphyllum fragilifolium* Dixon; Present status: Synonym of *Platygyriella fragilifolia* (Dixon) W.R. Buck

Distribution in India: India Orientalis

1501. *Trachyphyllum jeyporense* Thér. & Dixon; Present status: **Doubtful name**

Distribution in India: Central India (Endemic to India)

310. *Tuerckheimia* Broth.

1502. *Tuerckheimia svihlae* (E.B. Bartram) R.H. Zander; **Present status: Valid name**

Distribution in India: Eastern Himalayas

311. *Weissia* Hedw.

1503. *Weissia controversa* Hedw.; **Present status: Valid name**

Distribution in India: Western Himalayas and South India

1504. *Weissia ghatensis* Dixon & P. de la Varde; **Present status: Doubtful**

Distribution in India: Central India and South India

1505. *Weissia macrospora* Dixon & P. de la Varde; **Present status: Valid name**

Distribution in India: South India

1506. *Weissia rutilans* (Hedw.) Lindb.; **Present status: Valid name**

Distribution in India: Western Himalayas

1507. *Weissia wimmeriana* (Sendtn.) Bruch & Schimp.; **Present status: Valid name**

Distribution in India: Western Himalayas

Family: Calymperaceae Kindb.

312. *Octoblepharum* Hedw.

1508. *Octoblepharum albidum* Hedw.; **Present status: Valid name**

Distribution in India: Western Himalayas, Eastern Himalayas, South India, Central India and gangetic plains

313. *Syrrhopodon* Schwägr.

1509. *Syrrhopodon assamicus* H. Rob.; Present status: Synonym of *Syrrhopodon semiliber* (Mitt.) Besch.

Distribution in India: Eastern Himalayas

1510. *Syrrhopodon burmensis* (Hamp.) Reese & B. C. Tan; **Present status: Valid name**

Distribution in India: Eastern Himalayas

1511. *Syrrhopodon calymperoides* Cardot & P. de la Varde; **Present status: Doubtful**

Distribution in India: South India

1512. *Syrrhopodon gardneri* (Hook.) Schwägr.; **Present status: Valid name**

Distribution in India: Western Himalayas, Eastern Himalayas and South India

1513. *Syrrhopodon himalayanus* Dixon; Present status: **Synonym of** *Syrrhopodon tristichus* Nees & Schwägr.

Distribution in India: Western Himalayas

1514. *Syrrhopodon larmintaii* Broth. & Paris; **Present status: Valid name**

Distribution in India: Eastern Himalayas

1515. *Syrrhopodon leucophanoides* Cardot & P. de la Varde; **Present status:** Synonym of *Syrrhopodon prolifer* Schwägr.

Distribution in India: South India

1516. *Syrrhopodon semiliber* (Mitt.) Besch.; **Present status: Valid name**

Distribution in India: South India

1517. *Syrrhopodon spiculosus* Hook. & Grev.; **Present status: Valid name**

Distribution in India: Western Himalayas

1518. *Syrrhopodon strictus* Thwaites & Mitt.; **Present status:** Synonym of *Syrrhopodon gardneri* (Hook.) Schwägr.

Distribution in India: South India

1519. *Syrrhopodon subconfertus* Broth.; **Present status:** Synonym of *Syrrhopodon confertus* Sande Lac.

Distribution in India: Andaman Islands

P. ORDER: RHIZOGONIALES (M.FLESCH.) GOFFINET & W.R.BUCK.

Family: Rhizogoniaceae Broth.

314. *Rhizogonium* Brid.

1520. *Rhizogonium spiniforme* (Hedw.) Bruch; **Present status: Valid name**

Distribution in India: Western Himalayas, Eastern Himalayas and South India

Family: Racopilaceae Kindb.

315. *Racopilum* P. Beauv.

1521. *Racopilum cuspidigerum* (Schwägr.) Ångström; **Present status: Valid name**

Distribution in India: South India

1522. *Racopilum orthocarpum* Wilson & Mitt.; **Present status: Valid name**

Distribution in India: Western Himalayas, Eastern Himalayas and South India

1523. *Racopilum schmidii* (Müll. Hal.) Mitt.; **Present status: Valid name**

Distribution in India: South India

Q. ORDER: SELIGERIALES (M.FLESCH.) GOFFINET & W.R.BUCK.

Family: Seligeriaceae Schimp.

316. *Blindia* Bruch & Schimp.

1524. *Blindia campylopodiodes* Dixon & Badhw.; **Present status: Valid name**

Distribution in India: Western Himalayas

1525. *Blindia himalayana* Dixon & Badhw.; **Present status: Valid name**

Distribution in India: Western Himalayas

1526. *Blindia roerichii* R. S. Williams ; **Present status: Valid name**

Distribution in India: Western Himalayas

1527. *Blindia sordida* (Wilson & Mitt.) Müll. Hal.; Present status: **Synonym of** *Dicranodontium sordidum* (Wilson & Mitt.) Müll. Hal.

Distribution in India: Western Himalayas and Eastern Himalayas

R. ORDER: SPHAGNALES LIMPR.

Family: Sphagnaceae Dum.

317. *Sphagnum* L.

1528. *Sphagnum acutifolioides* Warnst.; **Present status:** Synonym of *Sphagnum junghuhnianum* Dozy & Molk.

Distribution in India: Eastern Himalayas

1529. *Sphagnum beccarii* Hampe; **Present status: Valid name**

Distribution in India: Eastern Himalayas

1530. *Sphagnum ceylonicum* Mitt. *ex* Warnst.; **Present status: Valid name**

Distribution in India: South India

1531. *Sphagnum cuspidatulum* Müll. Hal.; **Present status: Valid name**

Distribution in India: Eastern Himalayas

1532. *Sphagnum cuspidatum* Ehrh. ex Hoffm.; **Present status: Valid name**

Distribution in India: Eastern Himalayas

1533. *Sphagnum fimbriatum* Wilson; **Present status: Valid name**

Distribution in India: Western Himalayas

1534. *Sphagnum girgensohnii* Russow; **Present status: Valid name**

Distribution in India: Western Himalayas

1535. *Sphagnum imbricatum* Hornsch. ex Russow; **Present status: Valid name**

Distribution in India: Eastern Himalayas

1536. *Sphagnum junghuhnianum* Dozy & Molk.; **Present status: Valid name**

Distribution in India: Eastern Himalayas

1537. *Sphagnum khasianum* Mitt.; **Present status: Valid name**

Distribution in India: Eastern Himalayas

1538. *Sphagnum magellanicum* Brid.; **Present status: Valid name**

Distribution in India: Eastern Himalayas

1539. *Sphagnum nemoreum* Scop.; **Present status: Valid name**

Distribution in India: Eastern Himalayas

1540. *Sphagnum ovatum* Hampe; **Present status: Valid name**

Distribution in India: Eastern Himalayas

1541. *Sphagnum palustre* L.; **Present status: Valid name**

Distribution in India: Eastern Himalayas

1542. *Sphagnum papillosum* Lindb.; **Present status: Valid name**

Distribution in India: Eastern Himalayas

1543. *Sphagnum plumulosum* Roell.; **Present status: Valid name**

Distribution in India: Eastern Himalayas

1544. *Sphagnum recurvum* P. Beauv.; **Present status: Valid name**

Distribution in India: Eastern Himalayas

1545. *Sphagnum squarrosum* Crome ; **Present status: Valid name**

Distribution in India: Western Himalayas and Eastern Himalayas

1546. *Sphagnum subsecundum* Nees; **Present status: Valid name**

Distribution in India: Eastern Himalayas

1547. *Sphagnum teres* (Schimp.) Ångström; **Present status: Valid name**

Distribution in India: Eastern Himalayas

S. ORDER: SPLACHNALES GREV. & ARN.

Family: Meesiaceae Schimp.

318. *Amblyodon* Bruch & Schimp.

1548. *Amblyodon dealbatus* (Sw. & Hedw.) Bruch & Schimp.; **Present status: Valid name**

Distribution in India: Western Himalayas

319. *Meesia* Hedw.

1549. *Meesia triquetra* (L. ex Jolycl.) Ångström; **Present status: Valid name**

Distribution in india: South India

1550. *Meesia uliginosa* Hedw.; **Present status: Valid name**

Distribution in india: Eastern Himalayas

Family : Splachnaceae Grev.

320. *Gymnostomiella* M. Fleisch.

1551. *Gymnostomiella vernicosa* (Hook. & Harv.) M. Fleisch.; **Present status: Valid name**

Distribution in india: Western Himalayas, South India, Central India, Gangetic plains and Rajasthan

321. *Splachnobryum* Müll. Hal.

1552. *Splachnobryum bengalense* Gangulee; **Present status:** Synonym of *Kopononobryum bengalense* (Gangulee) Arts

Distribution in india: Gangetic plains and Rajasthan

1553. *Splachnobryum flaccidum* (Harv.) Braithw.; **Present status:** Synonym of *Splachnobryum obtusum* (Brid.) Müll. Hal.

Distribution in india: Gangetic plains and Rajasthan

1554. *Splachnobryum indicum* Hampe & Müll. Hal.; Present status: **Synonym of** *Splachnobryum obtusum* (Brid.) Müll. Hal.

Distribution in india: Western Himalayas, South India, Central India, Gangetic plains and Rajasthan

1555. *Splachnobryum procerrimum* Dixon & P. de la Varde ; Present status: **Synonym of** *Splachnobryum aquaticum* Müll. Hal.

Distribution in india: Western Himalayas and Central India

1556. *Splachnobryum synoicum* H. Rob.; Present status: **Synonym of** *Splachnobryum assamicum* Dixon

Distribution in india: Western Himalayas and Eastern Himalayas

322. *Tayloria* Hook.

1557. *Tayloria froelichiana* (Hedw.) Mitt. & Broth.; **Present status: Valid name**

Distribution in india: Western Himalayas

1558. *Tayloria imbricata* Thwaites & Mitt.**;** Present status: **Synonym of** Synonym of *Tayloria indica* Mitt.

Distribution in india: South India

1559. *Tayloria indica* Mitt.; **Present status: Valid name**

Distribution in india: Western Himalayas and Eastern Himalayas

1560. *Tayloria jacquemontii* (Bruch & Schimp.) Mitt.; **Present status: Valid name**

Distribution in india: Western Himalayas

1561. *Tayloria subglabra* (Griff.) Mitt.; **Present status: Valid name**

Distribution in india: Eastern Himalayas and South India

1562. *Tayloria tenella* Mitt.; Present status: **Synonym of** *Tayloria hornschuchii* (Grev. & Arn.) Broth.

Distribution in india: Western Himalayas

323. *Tetraplodon* B.S.G.

1563. *Tetraplodon angustatus* (Hedw.) Bruch & Schimp.; **Present status: Valid name**

Distribution in india: Eastern Himalayas

324. *Voitia* Hornsch.

1564. *Voitia hookeri* Mitt.; Present status: **Synonym of** *Voitia nivalis* Hornsch.

Distribution in india: Eastern Himalayas

T. ORDER: TETRAPHIDALES M. FLEISCH.

Family: Buxbaumiaceae Schimp.

325. *Buxbaumia* Hedw.

1565. *Buxbaumia himalayensis* Udar, S.C. Srivast. & D. Kumar ; Present **status:** Doubtful name **Distribution in india:** Western Himalayas (Endemic to India)

326. *Theriotia* Cardot

1566. *Theriotia kashmirensis* H. Rob.; Present status: **Synonym of** *Diphyscium kashmirense* (H. Rob.) Magombo, Zacharia Lekodi

Distribution in India: Western Himalayas

1567. *Theriotia lorifolia* Cardot; Present status: **Synonym of** *Diphyscium lorifolium* (Cardot) Magombo, Zacharia Lekodi

Distribution in India: Western Himalayas

U. ORDER: TIMMIALES OCHYRA

Family: Timmiaceae Schimp.

326. *Timmia* Hedw.

1568. *Timmia austriaca* Hedw.; **Present status: Valid name**

Distribution in India: Western Himalayas

1569. *Timmia bavarica* Hessl.; Present status: **Synonym of** *Timmia megapolitana* subsp. *bavarica* (Hessl.) Brassard

Distribution in India: Western Himalayas

1570. *Timmia megapolotana* Hedw.; **Present status: Valid name**

Distribution in India: Western Himalayas

Family: Hypopterygiaceae Mitt.

327. *Cyathophorella* (Broth.) M. Fleisch.

1571. *Cyathophorella adiantum* (Griff.) M. Fleisch.; **Present status: Valid name**

Distribution in India: Western Himalayas and Eastern Himalayas

1572. *Cyathophorella anisodon* Dixon & Herzog; Present status: Synonym of *Cyathophorum hookerianum* (Griff.) Mitt.

Distribution in India: Eastern Himalayas (Endemic to India).

1573. *Cyathophorella burkillii* (Dixon) Broth.; **Present status: Valid name**

Distribution in India: Eastern Himalayas (Endemic to India).

1574. *Cyathophorella hookeriana* (Griff.) M. Fleisch.; **Present status: Valid name**

Distribution in India: Eastern Himalayas

1575. *Cyathophorella intermedia* (Mitt.) Broth.; **Present status: Valid name**

Distribution in India: Western Himalayas and Eastern Himalayas

1576. *Cyathophorella tonkinensis* (Broth. & Paris) Broth.; **Present status: Valid name**

Distribution in India: Western Himalayas

328. *Hypopterygium* Brid.

1577. *Hypopterygium flavolimbatum* Müll. Hal.; **Present status: Valid name**

Distribution in India: Western Himalayas and Eastern Himalayas

1578. *Hypopterygium tenellum* C. Muell.; **Present status: Valid name**

Distribution in India: South India

Discussion

The current compilation of the moss flora of India revealed the occurrence of total 1578 species of mosses which belong to 21 orders, under 66 families and 328 genera. Out of these 897 retained their valid status, while 437 species are now considered as a synonym and status of 244 species is still unresolved i.e. doubtful name. 130 taxa have been reported as endemic to India. The updated checklist of mosses of India reveals that the most diversified order is Hypnales with 28 families, followed by Dicranales (6 families); Pottiales and Bryales (4 families each). Order Timmiales, Splachnales and Rhizogoniales have 2 families each and rest of the orders are represented by single family only. Whereas, in terms of family, the most prominent family is Pottiaceae having 38 genera followed by Hypnaceae with 20 genera. Genera like *Fissidens* (72 spp.), *Brachythecium*

(38 spp.), *Pogonatum* (33 spp.), *Mnium* (25 spp.), *Calymperes* (23 spp.), *Brachymnium* (22 spp.), *Sphagnum* (21 spp.), *Barbula* (21 spp.), *Entodon* (20 spp.) and *Tortula* (15 spp.) are most diversified genera in India. This great diversity of mosses reveals the tremendous potential of India in terms of bryodiversity predominantly, the mosses.

REFERENCES

AHMAD, T. (2011). Geology of the Himalayan Mountain Range, with special reference to the western Himalaya. GEOLOGI 63: 142-147.

ALAM, A. (2013). Moss flora of Munsiyari (Uttarakhand), Western Himalayas, India. *Archive for Bryology* 161: 1-11.

ALAM, A., KUMAR, A. & SRIVASTAVA, S. C. (2007). *Jungermannia* (*Plectocolea*) *nilgiriensis* sp.nov. from Nilgiri Hills (Western Ghats) India. *Bulletin Botanical Survey of India* 49 (1-4): 219-224.

ALAM, A., SHARMA, V., SHARMA, S. C. & TRIPATHI A. (2012). Bryoflora of Munsiyari and Dharchula Tehsil of Pithoragarh, Uttarakhand, Western Himalayas, India. *Archive for Bryology* 140: 1–11.

ALAM, A., PANDEY, S., SINGH, V., SHARMA, S.C., & SHARMA, V. (2014). Moss flora of Mount Abu (Rajasthan), India: An updated checklist. *Tropical Plant Research* 1(1): 8–13

ALAM, A., SHARMA, V., SHARMA, S.C., & YADAV, S. (2013). *Hypnum plumaeforme* Wilson - New addition to the Bryoflora of Western Himalayas, India. *Archive for Bryology*, 170: 1-4.

ALAM, A., BEHERA, K. K., VATS, S. & IQBAL, M. (2013). A preliminary study on bryodiversity of Similipal Biosphere Reserve (Odisha), India. *Archive for Bryology* 157: 1-9.

ASTHANA, V. & SAHU, V. (2013). Bryophyte Diversity in Mukteshwar (Uttarakhand): an overview. *Archive for Bryology* 154: 1-11.

AZIZ, M. N. AND VOHRA, J. N. (2008). Pottiaceae (Musci) of India, Bishen Singh ANDMahendra Pal Singh, Dehra Dun, India. pp. 366.

BUCK, W. R. AND GOFFINET, B. (2000). Morphology and classification of mosses. *In* A.J. Shaw & B. Goffinet (eds.), Bryophyte Biology. Cambridge University Press.: 71– 119.

CHOPRA, R. S. & KUMAR, S. S. (1981). Mosses of the western Himalaya. *Annales Cryptogamici et phytopathologici*, vol 5. The Chronica Botanica Co. New Delhi, India.

CHOPRA, R. S. (1975). Taxonomy of Indian Mosses. C.S.I.R. Publication, New Delhi, India.

DANIELS, A.E.D.; KARIYAPPA, K.C. (2013) *Fissidens angustifolius* (Fissidentaceae) - New to India from the Western Ghats. *Evansia* 30:76–78.

DABHADE, G.T. (1969) Mosses of Mahabaleshwar. *M.V.M. Patrika* 4(2): 94-104.

DABHADE, G.T. (1998) Mosses of Khandala and Mahabaleshwar in Western Ghats (India). Thane, India.

DANDOTIYA, D., GOVINDAPYARI. H., SUMAN, S., UNIYAL, P.L. (2011) Checklist of the bryophytes of India. *Archive for Bryology* 88: 1-126.

DANIELS, A.E.D. (2010) Checklist of the bryophytes of Tamil Nadu, India. *Archive for Bryology* 65: 1-117.

DANIELS, A.E.D., DANIEL, P. (2007) The mosses of the Southern Western Ghats. In: Nath V, Asthana AK, editors. Current trends in Bryology. Bishen Singh Mahendra Pal Singh, Dehra Dun, India, pp. 227-243.

DEORA, G.S., CHAUDHARY, B.L. (1996) Occurrence of *Bryum* Hedw. in Rajasthan. *Phytomorphology* 46: 299-304.

DIXON, H.N. (1909) Mosses from Western Ghats. *Journal of Botany, London* 47: 157-164.

GANGULEE, H. C. (1969-1980). Mosses of Eastern India and Adajacent regions. Fascicles 1-8. Books and Allied Limited, Calcutta.

GLIME, J. M. (2007). Bryophyte Ecology. Vol. I. Physiological Ecology. Ebook sponsored by Michigan Technological University and IAB.

GLIME, J. M., AND KNOOP, B. C. (1986). Spore germination and protonemal development of *Fontinalis squamosa. Journal of Hattori Botanical Laboratory.* 61: 487-497.

KISHWAN, J., PANDEY, R. AND DADHWAL, V. K. (2009). India's forest and tree cover: Contribution as a carbon sink. Technical Paper. Indian Council of Forestry Research and Education, PO: New Forest, Dehradun – 248006 (Uttarakhand) India

KUMAR, G.V., KRISHNAMURTHY, K.V. (2007) Moss flora of Shervaroy hills of Eastern Ghats (South India). In: Nath V, Asthana AK, editors. Current trends in Bryology. Bishen Singh Mahendra Pal Singh, Dehra Dun, India, pp. 37-45.

LAL, J. (2005). A checklist of Indian Mosses, Bishen Singh Mahendra Pal Singh. Dehra Dun, India. pp. 1–164.

LAKSHMINARAYANA, K. V; YAZADANI, G. M. AND RADHA-KRISHANAN. (2001). Western Ghats. *In Ecosystems of India.* (ed. Alfred, J. R. B.; Das, A. K. and Sanyal, A. K.). Pp. 349-369. ENVIS Centre, Zoological Survey of India, Kolkata.

MADHUSOODANAN, P.V., NAIR, M.C. (2004) Pleurocarpous mosses of Eravikulam National Park, Kerala-1. *Journal of Economic and Taxonomic Botany* 28: 338-346.

MADHUSOODANAN, P.V., NAIR, M.C., EASA, P.S. (2007) Diversity of bryophytes in Eravikulam National Park, Kerala (South India). In: Nath V, Asthana AK, editors. Current trends in Bryology. Bishen Singh Mahendra Pal Singh, Dehra Dun, India, pp. 255-267.

MANJU, C.N., RAJESH, K.P., JITHA, S., RESHMA, P.K. AND PRAKASHKUMAR, R. (2011). Bryophyte diversity of Kakkavayal Reserve Forest in the Western Ghats, Kerala. *Archive for Bryology* 108: 1-7.

NAIR, M.C. AND MADHUSOODANAN, P.V. (2006). A preliminary survey of the Bryophyte flora of Vellarimala in Western Ghats of Kerala. *Indian Journal of Forestry* 29(2): 191-196.

OXFORD (2014). Oxford School Atlas. Published by Oxford University Press-New Delhi.

PROCTOR, M. C. F. 1984. Structure and ecological adaptation. In: Dyer, A. F. and Duckett, J. G. (eds.). The Experimental Biology of Bryophytes. Academic Press, London, pp. 9-37.

RAJESH K.P., MUFEED, B. AND MANJU, C.N. (2013). *Symphyodon complanatus* (Symphyodontaceae: Moss) a new record for Kerala (India). *Samagra* 9: 28-30.

RAJESH, K.P. AND MANJU C.N. (2014). Bryophyte diversity of the lowlands and midlands of Kozhikode district, Kerala, India. *Frahmia* 5: 1-10.

RAVINDRANATH, N. H. MURTHY, I. K., JOSHI P, UPGUPTA, S, MEHRA, S. AND SRIVASTAVA, N. (2014). Forest area estimation and reporting: implications for conservation, management and REDD+. *Current Science*. 106 (9,10): 1201-1206.

NATH, V., ASTHANA, A. K. & SAHU, V. (2007). *Fabronia secunda* Mont. – A New addition to western Himalayas. *Indian Journal of Forestry* 30(3): 353-354.

NATH, V., ASTHANA, A. K. & SAHU, V. (2008). Addition of three moss species to West Himalayan Bryoflora. *Cryptogamie Bryologie* 29(4): 387-392.

SAXENA, D. & ARFEEN, M. S. (2009). Taxonomy and distribution status of moss *Racomitrium crispulum* in Kumaon hill of Western Himalaya (India)". *Iranian Journal of Botany* 15(2): 248–256.

SAXENA, D. & GANGWAR, R. (2005). Taxonomical study of *Dicranum scoparium* Hedw. from Kumaon hills. *Geophytology* 35 (1& 2). 61–64.

SAXENA, D., SINGH, S & SRIVASTAVA, K. (2006). Distribution of Some Mosses in Nainital, Almora and Pithoragarh District of Kumaon Region, India. *Environment Conservation* 7: (1–2): 83–87.

SAXENA, D., SINGH, S & SRIVASTAVA, K. (2007). Taxonomy of moss *Isopterygium elegans* (Brid.) Lindb., Not Sallsk. F. Fl Fenn. Forh., 1874 from Kumaon hills". Research Journal (Sci.), *Panjab University Research Journal (Sci.)*. 57: 213–216.

SAXENA, D., SONYIA & SUARABH. (2010). First report of the moss *Rhynchostegiella divaricatifolia* (Renauld & Cardot) Broth. from Western Himalayan region of India, *Phtyospecies*, U.K. **8**: 59–64.

SHAW, A. J. & GOFFINET, B. (2000). Bryophyte Biology. Cambridge University Press. 476 pp.

SINGH, H., V. SAHU, HUSAIN T. & ASTHANA, A.K. (2010). *Macromitrium rigbyanum* Dixon. In: Bryological Notes: New national and regional bryophyte records, 24. *Journal of Bryology* 32: 232–241.

SCHOFIELD, W. B. 1985. Introduction to Bryology. Macmillan Publishing Co., New York, 431 pp.

SCHWARZ, U. (2013). An Updated Checklist of Bryophytes of Karnataka. *Archive of Bryology* 181:1-42.

SCHWARZ, U.; FRAHM, J.-P. (2014) A contribution to the bryoflora of the Western Ghats in Karnataka State, India. *Polish Botanical Journal*. 58(2): 511-524.

TEWARI, S. D. & PANT, G. (1994). Bryophytes of Kumaon Himalaya. Bishen Singh Mahendra Pal Singh. Dehradun, India

VALDIYA, K.S. (1980). Geology of Kumaun Lesser Himalaya. Wadia Institute of Himalayan Geology, Dehradun, 291p.

VITT, D. H. 1984. Classification of the Bryopsida. In: Schuster, R. M. (ed.). New Manual of Bryology, Vol. 2. The Hattori Botanical Laboratory, Nichinan, Japan, pp. 696-759.

VERMA, P. K., ALAM, A., SRIVASTAVA, S.C. (2011) Status of mosses in Nilgiri Hills (Western Ghats), India. *Archive of Bryology* 102: 1-16.

VOHRA, J. N. (1970). A contribution to the moss flora of Western Himalaya. II. *Bulletin Botanical Survey of India* 12: 97–103.

Websites:

www.theplantlist.org

www.envindia.com

www.excellup.com

www.ncert.nic.in

www.wikipedia.org

www.importantindia.com